THE LATE MINOAN III NECROPOLIS
OF ARMENOI, VOLUME II

John Martlew

Angeliki Tzedaki

We lost two outstanding and long-term supporters, without whose love and encouragement the project would have been sadly diminished. It is to their memories that Volume II is dedicated.

THE LATE MINOAN III NECROPOLIS OF ARMENOI, VOLUME II

BIOMOLECULAR AND EPIGRAPHICAL INVESTIGATIONS

Edited by

YANNIS TZEDAKIS, HOLLEY MARTLEW
AND MICHAEL TITE

with contributions by

PETER W. DITCHFIELD, CEIRIDWEN J. EDWARDS,
M. GEORGE B. FOODY, LOUIS GODART, VICKY KOLIVAKI,
OLGA KRZYSZKOWSKA, MICHAEL P. RICHARDS,
MAGDALENA WACHNIK AND DARLENE WESTON

OXBOW | books
Oxford & Philadelphia

Published in the United Kingdom in 2024 by
OXBOW BOOKS
The Old Music Hall, 106–108 Cowley Road, Oxford, OX4 1JE

and in the United States by
OXBOW BOOKS
1950 Lawrence Road, Havertown, PA 19083

Paperback Edition: ISBN 979-8-88857-046-3
Digital Edition: ISBN 979-8-88857-047-0 (epub)

A CIP record for this book is available from the British Library

Library of Congress Control Number: 2023948964

Printed in the United Kingdom by Short Run Press

Typeset in India by Lapiz Digital Services, Chennai.

For a complete list of Oxbow titles, please contact:

UNITED KINGDOM
Oxbow Books
Telephone (0)1226 734350
Email: oxbow@oxbowbooks.com
www.oxbowbooks.com

UNITED STATES OF AMERICA
Oxbow Books
Telephone (610) 853-9131, Fax (610) 853-9146
Email: queries@casemateacademic.com
www.casemateacademic.com/oxbow

Oxbow Books is part of the Casemate Group

Front cover: Larnax RM 1710, Tomb 10 LM IIIA:2-IIIB:1
Back cover: Selection of Larnakes (top row): R.M. 1707, Tomb 11; R.M. 5121, Tomb 139; R.M.1712, Tomb 24; R.M. 1703, Tomb 55; (bottom row): RM 1703, Tomb 55; R.M. 1709, Tomb 10; R.M. 1706, Tomb 24; R.M. 2334, Tomb 18 (photos by Stefanos Alexandrou)

Contents

List of figures and tables

Figures

Tables

Contributors

YANNIS TZEDAKIS
Samara, 27, Paleo Psychico, 15452 Athens, Greece

HOLLEY MARTLEW
The Hellenic Archaeological Foundation, Tivoli House, Tivoli Road, Cheltenham GL50 2TD, UK

MICHAEL TITE
Research Laboratory for Archaeology, School of Archaeology, University of Oxford, 1 Parks Road, Oxford OX1 3TG, UK

PETER W. DITCHFIELD
Research Laboratory for Archaeology, School of Archaeology, University of Oxford, 1 Parks Road, Oxford OX1 3TG, UK

CEIRIDWEN J. EDWARDS
Department of Biological and Geographical Sciences, School of Applied Sciences, University of Huddersfield, Queensgate, Huddersfield HD1 3DH, UK

M. GEORGE B. FOODY
Department of Biological and Geographical Sciences, School of Applied Sciences, University of Huddersfield, Queensgate, Huddersfield HD1 3DH, UK

LOUIS GODART
Viale Liegi 48C, 00198 Rome, Italy

VICKY KOLIVAKI
VK Studio of Architecture and Restoration, Sifi Vlastou 14, 74100 Rethymnon, Crete, Greece

OLGA KRZYSZKOWSKA
Institute of Classical Studies, University of London, Senate House, Malet Street, London WC1E 7HU, UK

MICHAEL P. RICHARDS
Department of Archaeology, Simon Fraser University, Burnaby, British Columbia, Canada V5A 1S6

MAGDALENA WACHNIK
73 Minchery Road, Oxford, OX4 4LU, UK

DARLENE WESTON
Department of Anthropology, University of British Columbia, 2104–6303 NW Marine Drive, Vancouver, British Columbia, Canada V6T 1Z1

The Armenoi Project

Directors

Yannis Tzedakis
Holley Martlew

Excavator

Yannis Tzedakis

Late Minoan III Necropolis of Armenoi: specialists

Pottery: Yannis Tzedakis and Vicky Kolivaki
Bronzes: Ioanna Efstathiou
Larnakes: Katarina Baxivani
Small finds: Vicky Kolivaki
Seals: Olga Krzyszkowska
Dromoi: Holley Martlew and Vicky Kolivaki
Stele: Eleni Papadopoulou
Miniature vases: Anna Simandiraki
Catalogue entries: Vicky Kolivaki
Skeletal material: Robert Arnott
Osteoarchaeologist: Darlene Weston
Geologist: †Andrew P. Giże
Head Surveyor: †Stephen Litherland

Archaeological illustrator: Magdalena Wachnik
Archivist (Greece): Vicky Kolivaki
Archivist (UK): Michael Jones

Consultant archaeologists

Irini Gavrilaki
Michael P. Richards
Olga Krzsyzkowska

Medical consultant

Dimitri Michaelides

IT consultants

Charalambus Alexandrou (Greece)
Mark Scarborough (UK)

Site guardians

Charalambus Litanas
Eugenios Psiharakis

Acknowledgements

Without the dedicated and inspired work of Yannis Tzedakis and his belief in me, none of our work, including the innovative scientific project reported here would have taken place. It was the expertise and hard work of the contributors who made publication of this volume possible. The expertise and dedication of Michael Tite as Scientific Editor was also an essential component for the realisation of Volume II. My appreciation to the contributors and to Michael Tite knows no bounds. I wish to thank Robert Arnott for the huge amount of time and effort he expended during the compilation of the book. I wish to thank Mark Scarborough, Jess Hawxwell, and Julie Gardner who helped immeasurably in the final preparation of the images and text, and did so with great patience and skill. I wish to thank Olga Krzyszkowska for her advice and support. Vicky Kolivaki supported the editors in their every need. She diligently compiled the catalogue entries and co-directed the excavation of the 'city' of Armenoi. Magdalena Wachnik is an archaeological illustrator par excellence. Her contribution to Volume II was invaluable.

The Field Survey was directed by Eileen Chappell and Steve Allender. We have them and their team to thank for the groundwork that led to the discovery of the 'city' of Armenoi. Thank you to our consultants and archivists, especially Irini Gavrilaki, for the time she gave, and to all members of the Armenoi Project.

All those whose help we acknowledged in Volume I, we thank again here, most especially The Headley Trust, INSTAP, the Holley Martlew Archaeological Foundation and the Hellenic Archaeological Foundation. We also thank two new friends, Virginia Giannitsas and George Georgakakis.

All of us wish to give special thanks to Jess Hawxwell and Julie Gardiner, the editors at Oxbow Books, for their efficiency and enthusiasm.

As for me … when the lights were shining brightly all of us were very grateful. When the lights went out, I never gave up. Index and additional expenses incurred, courtesy of Holley Martlew.

We remember those who worked with the project and are no longer with us. During the time this volume was being prepared, we witnessed the untimely deaths of our head surveyor, Stephen Litherland and our geologist, Andrew P. Giże. They will never be forgotten.

Holley Martlew

The results of scientific analyses allow all of us to identify with our ancestors in a way nothing else can do. Much the same can be said for cracking the codes of an ancient language, in this case Linear B. Revealing the humanity of our ancestors, what they ate and drank, even who their relatives were is undoubtedly the reason why scientific analyses such as staple isotopes and ancient DNA have become so popular with scientists and laymen alike, and why their fascination will not wane but wax in perpetuity. These are the reasons why the editors of the series on the Late Minoan III Necropolis of Armenoi will forever be in debt to the scientists and to a renowned Linear B expert, who produced the outstanding results published in Volume II. We thank them with all our hearts for their efforts. Thank goodness the Necropolis lived up to the task.

Yannis Tzedakis and Holley Martlew

Conventions, abbreviations and chronology

Abbreviations

The standard terminologies are those in use within the academic fields of Aegean prehistory, classical archaeology and geology. Specific abbreviations in use in the volume are:

Cat. No.	Catalogue number
cm	Centimeters
CMS	*Corpus der minoischen und mykenischen Siegel*
D	Diameter
FM	Furumark motif
FS.	Furumark shape
H	Height
km	Kilometres
L.	Length
M.	Metal find
m.	Metres
max.	Maximum
min.	Minimum
mm.	Millimetres
MNI	Minimum Number of Individuals (skeletons found in tomb)
O	Ivory or bone find
Pres. max. dim.	Preserved maximum dimensions
Σ	Seals
Th.	Thickness
W.	Width
Y	Glass find

Museums

R.M.	Archaeological Museum of Rethymnon
Ch.M.	Archaeological Museum of Chania
H.M.	Archaeological Museum of Heraklion
M.N.	Archaeological Museum of Nauplia

Chronology

The chronology, both relative and absolute that the editors have followed is based on Peter Warren and Vronwy Hankey (1989), *Aegean Bronze Age Chronology*. Bristol, Bristol Classical Press although a slightly revised chronology for the Late Minoan III period as follows, is in use by the authors:

Late Minoan IIIA:1 1390–1370 BC
Late Minoan IIIA:2 1370–1340 BC
Late Minoan IIIB:1 1340–1250 BC
Late Minoan IIIB:2 1250–1190 BC
Late Minoan IIIC   1190–1130 BC

Foreword

Retention and preservation of skeletal material

The excavation of the Late Minoan III Necropolis of Armenoi stands nearly alone historically for its retention and conservation of skeletal material from Minoan and Mycenaean burials, be they individual or in a cemetery. It is the great good fortune of Minoan and Mycenaean studies that the Necropolis is large, intact and rich in both artefacts and skeletal material. To date 232 tombs have been excavated.

As co-director of the scientific project initiated in 1997, which culminated in an Exhibition mounted in six international museums, Holley Martlew visited the storerooms of excavations in Crete and on the Mainland, to take samples of ceramic artefacts and skeletal material. The retention of ceramic artefacts from excavations goes without saying, but keeping human skeletal material was apparently another thing as Martlew discovered in the mid-1990s. She travelled throughout Mainland Greece, inquiring after and searching for skeletal material from old excavations. In most instances it was not there, nothing had been conserved. It was at that point that she became aware of the extent to which Tzedakis's excavation at the Necropolis was becoming the benchmark for archaeology. By systematically retaining the skeletal material, tomb by tomb since the day he started excavating, Tzedakis had been a pioneer.

The sites from which Martlew was able to obtain a sufficient number of samples so that Stable Isotope Analysis on human skeletal material could be carried out, were the Neolithic Cave of Gerani (Tzedakis excavation), Grave Circles A (Schliemann 1876) and B (Papadimitriou 1952–54) at Mycenae, a group of chamber tombs at Mycenae (Eleni Palaeologou), and the Late Minoan III Necropolis of Armenoi itself (Tzedakis excavation).

The scientific project initiated in 2017 focused on the conserved skeletal remains from Necropolis of Armenoi to carry out DNA sequencing, stable isotope analysis, osteological analysis, and radiocarbon dating, This policy of systematic retrieval, retention and conservation means that scientists now have access to the remains of about 1000 people from a magnificent Necropolis which, with a time span of 200 years, is a microcosm of the Minoan world in Late Minoan III.

There are three other factors to thank: (a) Foresight: we have always been in pursuit of a subject best described as Archaeology meets Science; (b) Environmental conditions, which helped the DNA and collagen to survive in the skeletal material; and (c) Luck, which one always needs to succeed at anything.

Plan of the Late Minoan III Necropolis of Armenoi with all tombs numbered

The Late Minoan III Necropolis of Armenoi: introduction

Holley Martlew

High above Rethymnon in West Crete, the Late Minoan III Necropolis of Armenoi (ca. 1390–1190 BC) sits on a gently sloping hillside. The site is 9 km south of the town. It covers 3 ha of hillside at 355–365 m above sea level (Figs 1.1 and 1.2). It is a complete cemetery. This alone sets it apart: it is the only known intact cemetery of the Late Minoan period. Because the 232 chamber tombs excavated to date held a treasure trove of artefacts and skeletal remains, and the Necropolis dates to a finite, 200 year period, it is so far unique. The Late Minoan III Necropolis of Armenoi is a microcosm of the Late Minoan world at its finest.

The excavated chamber tombs are virtually unplundered, a rarity for any archaeological site. The abundance and the artistic excellence of the artefacts is extraordinary and, throughout the years of excavation, its excavator, Yannis Tzedakis, unusually for the time, had all the human skeletal material retained and conserved. This allowed for scientific research to be carried out which, in turn, has had outstanding results, which are presented in this volume.

In terms of artefacts, the Necropolis delivered in abundance. There have been exceptional finds such as incense burners painted with flowers and elegant, stemmed goblets, many of which have bowls decorated with octopi whose tentacles dangle daintily down the long stems. Goblets were tin-plated to imitate the gleam of silver (see Fig. 9.28). There are vases and jugs painted with birds, flowers and sea creatures. A unique stirrup jar has a Linear B inscription painted on it (Fig. 9.38). Over 800 vases have been found. Bronze objects, of which more than 300 have been recovered, include a dipper (Fig. 9.25), tweezers (Fig. 9.24), a cleaver (Fig. 6.3), knives and swords. A reed basket decorated with bronze nails was found (Tzedakis *et al.* 2018, fig. 1.49) and a boar's tusk helmet (Fig. 9.26), one of only two found in Crete to date.

Thirty-three decorated larnakes/sarcophagi were found in the tomb chambers, most of them with gabled lids and side panels that imitate painted wooden chests. Ten are in pieces waiting to be restored (see Chapter 9). Motifs are Minoan religious symbols and activities, such as horns of consecration, double axes and a goddess with raised arms (illustrated on the back cover), and a sacred hunt – men in short tunics, arms raised and holding weapons, watch as javelins rain down, their points just making contact with the backs of wild goats (back cover). A procession of bulls with rosettes painted on their bodies is on another (front cover). There are painted rows of wide-eyed octopi with curling tentacles on several larnakes (back cover).

Originally the tombs were cut horizontally into an east facing, unwooded, white limestone hillside, and, according the geologist Andrew Giże, the tomb entrances would have shone in the distance like the white cliffs of Dover for a period of 200 years. This would have made quite a statement to the population at the time about the wealth, the power and the importance of the town and its rulers.

The chamber tombs were thought by the excavator to be dynastic and this has been proven through osteological and DNA analysis of skeletal material (Chapters 3 and 5). They vary in size, type and number of artefacts, and they differ in state of preservation because some of the roofs have fallen into the chambers and due to havoc caused by badgers. At the entrances many large stones had been propped up in the doorways to seal the tombs.

To the left of the entrance to the Necropolis, part way up the hill, is the magnificent Tomb 159, the largest and wealthiest burial place in the Necropolis (Chapters 3–6, 8, 9 and 10 postscript). It is presumed to have been built for a ruler. Because the tombs differ in size and design, they also differ in the effect each has on the observer. No matter how many times one has visited the Necropolis, it is an awesome experience to walk down the long, stepped corridor of Tomb 159 into the cavernous chamber and then to look back up the stairs into the sunlight (Fig. 9.34 gives an idea of the scale). By way of contrast, if one continues up the hill, there is Tomb 24 (Figs 9.35 and 9.36) whose

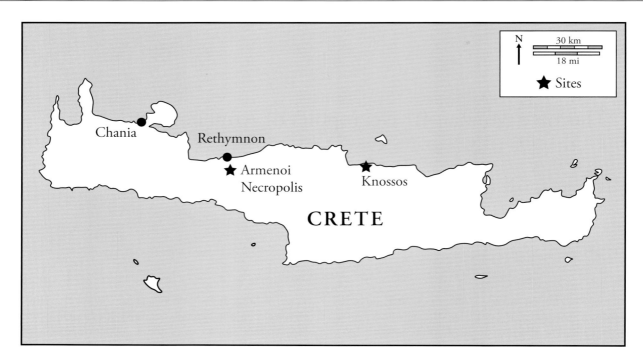

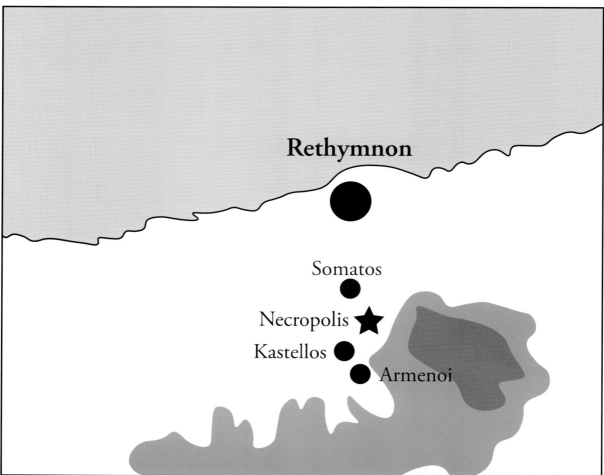

Figure 1.1. Location of the Late Minoan III Necropolis of Armenoi on Crete.

Figure 1.2. View of a section of the Necropolis from the west.

corridor is very narrow and cut deeply into the limestone. It has a totally different aspect. Going down into it, one feels as though descending into the bowels of the earth. It, like Tomb 159, was, however, a very wealthy tomb and, like Tomb 159, contained two painted larnakes.

The discovery of the Necropolis

One day in 1969 a young boy took his usual short cut from his home in the village of Somatas to his school in the village of Armenoi by diagonally crossing a low hill that sloped

below his family's (and now his) property. He found a vase which he took to the school and gave to the school master. The vase is illustrated in Figure 1.3. The schoolmaster took it to the Archaeological Museum in Chania where the painted decoration was immediately recognised as Late Minoan III. The guard at the museum telephoned Yannis Tzedakis, who was the Ephor (Director) of Archaeology in West Crete at the time and was working at an excavation along the coast near the village of Pigi. The next day Tzedakis took two experts to the site: Stavros Polychronakis, his chief workman and head of security at Chania Museum, and Spiros Vassilakis,

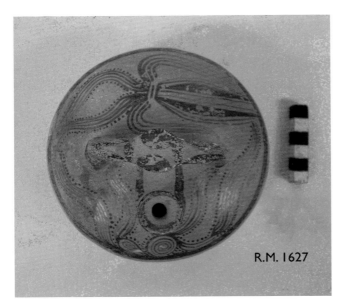

Figure 1.3. Juglet which led to the discovery of the Necropolis (R.M. 1627).

the chief technician for the British School at Athens at their Knossos excavations who, as a small boy, worked for Sir Arthur Evans at the Palace of Knossos. The men walked around slowly. They found pots and a few bones. They continued to study the terrain and after a while it became clear to Stavros that beneath their feet was definitely a Minoan necropolis, and a very large one. The men thought the only reasonable explanation was that badgers had made lairs in the tombs. In order to clean them out, they had tossed pots and bones up onto the ground above them.

A few days later Tzedakis brought a team of workmen to the spot and, under the auspices of the Greek Ministry of Culture, they started an excavation. It soon became obvious that the young boy had been instrumental in the discovery of one of the most important Minoan sites in Crete. The excavation lasts into the 21st century as does the careful conservation of material.

Method

In order to locate tombs, the chief workman used the time-honoured method of thumping the ground with a long, thick pole. He would listen intently for an echo which indicated there was a hollow area beneath the surface. Today there are more sophisticated methods that can be used. During the Field Surveys of 2001–2007 Eileen Chappell and Steve Allender used a Geoscan RM15 resistance meter.

At the time the Necropolis was discovered Tzedakis was Director of Antiquities, West Crete. He then became Director for Central and West Crete and subsequently the Minister of Culture in Athens to serve as Director of Antiquities. After that he became Director General of the Greek Archaeological Service (now Emeritus). The

importance of the Necropolis was such, however, that the Ministry allowed him to set aside time for excavation under their auspices nearly every year. The years that are omitted in the list below are years when Tzedakis was carrying out official duties for the Ministry of Culture, including 2001–2004 when he was Director of Culture for the 2004 Olympics.

The tombs

The tombs are numbered in the order they were discovered. Tzedakis believes that the 232 tombs that have been excavated to date represent the majority within the Necropolis but he is equally sure that there are a few tombs still to be discovered. It is unlikely that all the tombs will ever be located. In spite of the sophisticated equipment now available there is no way of ever being certain because of the difficult geology of the terrain. When time and funds permit, a search is made for more tombs but, at the present time, the directors and members of the Armenoi Project are expending most of their efforts to realise publications. Finds from the Necropolis are on permanent display in the Archaeological Museum of Rethymnon.

List of tombs and the years they were excavated

Year	tombs excavated
1969	1–8
1970	8–25
1971	26–64
1973	65–91
1976	92–107
1978	108–124
1980	125–142
1984	143–158
1985	159–167
1986	168–176
1987	177–182
1988	183–196
1989	197–207
1990	208–212
1991	213–217
1993	218–221
1995	222–223
1997	224–228
2001–present	229–232

Publication programme

The programme for publishing the results of the excavations is well underway. Volume I (Tzedakis *et al.* 2018) focused on the physical background and setting of the project; the field and geophysical surveys; and the history of the excavations, with presentations on the key assemblages of Minoan, Roman and Byzantine diagnostic pottery recovered by the surveys. A proposed method of tomb construction was set out. Volume I also discusses mineralogical investigation of an iron deposit that occurs approximately 5 km west of the Necropolis, on the eastern margin of Ano Valsamonero, as a possible source of metal used for objects present in the tombs, and introduces the question of identifying places mentioned in the Linear B Co series of tablets from Knossos.

Volume III will present an encyclopaedia of the tombs, tomb plans and catalogue entries with photographs of the most important finds, published in three parts. Part 1, tombs 1–75, is nearing completion. Volume III.2 catalogues Tombs 75–159 and Volume III.3, Tombs 160–232. Volume IV will concentrate on artefacts. The subjects are pottery, bronzes, seals, larnakes (sarcophagi), stele (tombstones), small finds, miniature vases and the contents of the dromoi (corridors which lead into the tombs), where ritual activity clearly took place.

The present volume focuses on the suite of scientific analyses undertaken on the skeletal remains, animal bones and ceramic vessels: organic residue, osteological, stable isotope and ancient DNA. Whilst, to date, only a small selection of materials has been subject to these studies the results are already remarkable and of international significance, providing the key first positive evidence for Minoan familial genealogy, population movement and the relationship between the diet of the local population and possible husbandry practices at a crucial period in the economic and political development of Bronze Age Crete and the eastern Mediterranean.

In 2003 we decided to focus on finding the 'city' which built the Necropolis. To this end we initiated field and geophysical surveys. Tzedakis consulted renowned Linear B expert, Louis Godart, and asked him what information might be contained in the Linear B tablets from Knossos that could relate to the Necropolis and the 'city'.

The scientific programme

The scientific programme was directed by the present author and Yannis Tzedakis. It began with a programme of organic residue analysis on ceramic artefacts to try to determine what the Minoans cooked, ate and drank (Chapter 2). After the programme commenced and it became clear that it was a success, Yannis Tzedakis and I decided that we should expand the project to include human skeletal material if such were possible. This discussion took place in Crete. Robert Hedges and Michael Tite of the Research Laboratory for Archaeology and Art History at the University of Oxford recommended stable isotope analysis for dietary analysis and proposed suitable specialists. The isotope work was undertaken by Michael Richards (see Chapter 4) and Darlene Richards was invited to carry out osteoological analysis on skeletal material (Chapter 3). The project was eventually expanded to include 16 sites including examples in Greece and the island of Vivara in the Bay of Naples and culminated in two books (Tzedakis *et al.* 2000; 2008), the second of which presented the primary scientific evidence behind each interpretation, and an exhibition. The latest programme of scientific work (isotope and DNA studies) was initiated and is reported on in Chapters 4, 5 and 8.

Bibliography

Tzedakis, Y. and Martlew, H. (eds) (2000) *Minoans and Mycenaeans: flavours of their time*. Athens, Ministry of Culture (Exhibition Catalogue).

Tzedakis, Y., Martlew, H. and Arnott, R. (eds) (2018) *The Late Minoan III Necropolis of Armenoi* Volume I. Philadelphia PA, INSTAP Academic Press.

Tzedakis, Y., Martlew, H. and Jones, M.K. (eds) (2008) *Archaeology Meets Science: biomolecular investigations in Bronze Age Greece*. Oxford, Oxbow Books.

Part I

Scientific analyses

Food and drink: what scientific analysis of pottery revealed through organic residue analysis

Holley Martlew

Introduction

Organic residue analysis was carried out on material from the Necropolis in the original project, directed by Tzedakis and Martlew, which pioneered the application of state of the art scientific analyses to ceramic artefacts and human skeletal material (1997 and 2003) at 16 sites in the Greek Mainland, Crete, other Greek islands, and the island of Vivara in the Bay of Naples (Tzedakis and Martlew 2003). The primary scientific results were published by Tzedakis *et al.* in 2008. The scientific results *in toto* were presented in an exhibition mounted in seven international museums (Athens, Rethymnon, Naples, Birmingham, Geneva, Stockholm (where it was opened by Queen Sonia), and Chicago (Tzedakis and Martlew 2000)). It was the most travelled of any exhibition the Greek government ever sent abroad.

In cases where the chemical signals survive, an experienced organic chemist is able to identify what foodstuffs and drink a vessel has contained. In the beginning (1997) we could only use sherds. By 2001, organic chemist Victor Garner had developed a non-destructive method which meant intact vessels (and soils found inside them) could be submitted for analysis. Intact vessels could not be sent out of Greece.

Materials and methods

A broken tripod cooking pot (R.M. 17313/2) found in the corridor of Tomb 177, gave a result of meat and oil. Another tripod (EUM 302) had been used to cook meat, cereal and pulses (possibly lentils) in olive oil. One would suppose that cooking pots would have been used for preparation of stews but it is another thing to be able to prove it scientifically, and likewise with the drinking vessels described below.

The remains of a tall stemmed tin-plated kylix/goblet (R.M. 17284) and a cup (R.M. 17337) revealed resinated wine, barley beer and honey mead. It was suggested that this strange combination could have been a Minoan cocktail comprised of all three. As tin-plated goblets imitate silver, it is noteworthy and fitting that such goblets would have been used to hold a 'royal' cocktail, as some of these vessels were found in the 'royal' Tomb 159 (Chapter 9) and others in ceremonial pits. The positive organic results for vessels from the Necropolis are listed below.

A compendium of methods was used to examine the residues: mass spectrometry; gas chromatography; infra-red spectroscopy; high performance liquid chromatography; scanning electron microscopy; wet chemical spot tests; XRF of inorganic elements in powder sample, IR and HPLC-UV analyses of organic residues obtained by sequential solvent extraction; and qualitative colorimetric detection of tartrate and calcium oxalate residues (Tzedakis and Martlew 2000; Tzedakis *et al.* 2008). These analyses were carried out by a team of leading specialists in the USA and England: †Curt W. Beck (Vassar College, NY, who was instrumental in helping me direct the programme and who carried out most of the work); Patrick E. McGovern (University of Pennsylvania); Ruth Beeston (Davidson College, NC); †John Evans (University of East London), †Victor Garner (Hall Analytic, Wythenshawe); Waters; MS Technologies, UK and †Andrew P. Gize (University of Manchester).

Results

The results for vessels from the Necropolis (taken from eight sherds and four soil samples) submitted for organic residue analysis are listed below. The numbers are from Tzedakis and Martlew (2003). asterisks denote vessels mentioned above. The organic residue results are in italics.

84.   Tripod cooking pot
       EUM 338. Armenoi 'city' Sector 11, wall/base
              sherd
       LM IIIA:1
       Olive oil
       Tzedakis and Martlew 2000, 112

86.   Juglet
       R.M. 22855. Tomb 226, soil sample
       LM IIIA:2
       *Copious amounts of plant waxes. No fats or oils ever
              held in this vessel.*
       Tzedakis and Martlew 2000, 113

87.   Amphora, soil sample
       R.M. 22852. Tomb 227
       LM IIIA:2
       Olive oil or plant oil
       Tzedakis and Martlew 2000, 113

89.   Cylindrical alabastron, soil sample
       R.M. 2676. Tomb 132, tomb with two skulls recon-
              structed for the Exhibition
       LM III
       Oil. Pulses/lentils
       Tzedakis and Martlew 2000, 114

*90.   Tripod cooking pot, soil sample
       R.M. 17313/2. Tomb 177
       LM IIIB
       Meat, olive oil
       Tzedakis and Martlew 2000, 115

91.   Tripod cooking pot, wall sherd
       R.M. 17322/1. South-west of Tombs 177/178
       LM IIIB
       Olive oil, milk?
       Tzedakis and Martlew 2000, 115

92.   Tripod cooking pot, wall sherd
       R.M. 17326/14. South-west of Tomb 178
       LM IIIB
       Olive oil, complex mixture
       Tzedakis and Martlew 2000, 116

*93.   Tripod cooking pot, wall sherd
       EUM 302. Tomb 178
       LM IIIB
       Traces of olive oil, cereal, meat pulses/lentils?
       Tzedakis and Martlew 2000, 116

94.   Tripod cooking pot, wall sherd
       EUM 303. Tomb 178
       LM IIIB
       Olive oil, complex mixture
       Tzedakis and Martlew 2000, 116

*167. Tin-plated kylix, wall sherd
       R.M. 17284. Ceremonial pit, Tomb 178
       LM IIIA:2
       *Resinated wine, barley beer and honey mead as part
              of mixed fermented beverage*
       Tzedakis and Martlew 2000, 176

168.   Jug, wall sherd
       R.M. 17340. Ceremonial pit, Tomb 178
       LM IIIA:2
       *Possibly resinated wine, barley beer and honey mead
              as part of mixed fermented beverage*
       Tzedakis and Martlew 2000, 176

*169. Cup, wall/base sherd
       R.M. 17337. Ceremonial pit, Tomb 178
       LM IIIA:2
       *Resinated wine, barley beer and honey mead as part
              of mixed fermented beverage*
       Tzedakis and Martlew 2000, 176

Bibliography

Garner, V. (2008) Alternative approaches to organic residue analysis: the Early Helladic Cemetery at Kalamaki, the Mycenaean settlement on Salamis; the Late Helladic Cemetery at Sykia; Vivara, settlement of Punta d'Alaca, Bay of Naples, Italy. In Tzedakis *et al.* (eds) 2008, 144–62.

Tzedakis, Y. and Martlew, H. (eds) (2003) *Minoans and Mycenaeans: flavours of their time* (revised ediion). Athens, Ministry of Culture.

Tzedakis, Y., Martlew, H. and Arnott, R. (eds) (2018) *The Late Minoan III Necropolis of Armenoi* Volume I. Philadelphia PA, INSTAP Academic.

Tzedakis, Y., Martlew, H. and Jones, M.K. (eds) (2008) *Archaeology Meets Science: biomolecular investigations in Bronze Age Greece.* Oxford, Oxbow Books.

The osteological study of Tomb 159

Darlene Weston

Introduction

The 232 excavated rock-cut family tombs of the Late Minoan III Necropolis of Armenoi contained the skeletal remains of approximately 1000 individuals. The Necropolis has yielded the largest single sample of human skeletal remains from the Greek Late Bronze Age known to date (Tzedakis and Martlew 1999; Tzedakis and Kolivaki 2018). Of the 1000 individuals ca. 200 were articulated primary burials and 800 were secondary co-mingled burials. The secondary burials were not true secondary burials but earlier interments pushed aside to make room for new tomb inhabitants. The variation in tomb size and in the wealth of associated grave goods suggests status differences may have been apparent between individuals and family groups (Papadopoulou 2017).

Tomb 159: its characteristics and contents

Tomb 159 is located among a group of 47 widely spaced tombs occupying the top of Prinokephalo Hill ('hill of the wild oaks'). This group is dominated by large tombs (>15 m²), has fewer small tombs (<10 m²), and has the most tombs with impressive architectural features. Among this group, Tomb 159 stands out and has been described as the finest example of funerary architecture in western Crete dating from the Late Minoan III period (Papadopoulou 2017).

Among Tomb 159's architectural features are a 15.5 m dromos with 25 steps descending to an entrance that mimics a house door. This doorway has an anathyrosis (double defining frame) with square bases on both sides, likely to support columns. This arrangement, in effect, would have created a kind of propylon, or monumental pillared gate, which would have signified the tomb belonged to a high prestige individual or family (Papadopoulou 2017). The tomb's chamber was approximately 20 m² (the largest on the site) and contained a niche carved into the wall, a pillar located directly opposite the door and a bench running the chamber's length (Tzedakis and Kolivaki 2018).

Inside the chamber were found the remains of a limestone stele (tombstone; R.M. 741) – one of 12 found on the site and one of only two found in direct association with a tomb (the other being Tomb 24). The stele was destroyed by robbers who looted the tomb at the end of the Minoan period, after the Necropolis had been abandoned. When the tomb was excavated in 1985 it was found to be closed with small stones (Tzedakis and Martlew 1999). The stele likely provided a signal for the robbers regarding the potential richness of the tomb's contents as, in Crete, grave markers were uncommon and reserved for the tombs of prominent persons (Papadopoulou 2017). Interestingly, it appears that, at Armenoi, only tombs with stele were robbed (Tzedakis and Martlew 1999; 2007).

Because Tomb 159 was looted, the richness of its contents will never entirely be known, but some artefacts were contained within including a larnax, jewellery and the remains of a wooden bier (Löwe 1996). In addition, several kylikes – wine-cups with religious or ceremonial significance (R.M. 3455, R.M. 3459, R.M. 3460 (Fig. 9.28)) were found in the tomb. Scientific analysis of kylikes at Armenoi has demonstrated they contained resinated wine, barley beer and mead (see Chapter 2; Tzedakis and Martlew 1999).

Based on its architecture and contents there seems little doubt that Tomb 159 belonged to a family of prominent individuals, but was that high status also reflected in the bones and teeth of the individuals buried within? To answer this question, osteological analysis of the individuals buried within Tomb 159 was undertaken with the goal of creating individual osteobiographies.

Burial context and MNI

It was not possible to discern the specific burial contexts of the individuals buried in Tomb 159 as the excavation drawings of the tomb did not survive, however, osteological analysis determined that the minimum number of individuals

(MNI) present in the tomb was five: 159A, 159B, 159Γ, 159Δ and 159E. MNI was determined based on the presence of four adult left humeri and six non-adult cranial bone fragments. Based on previous conventions for the reporting of individuals from the Armenoi tombs (Y. Tzedakis, pers. comm.), the last interred individual was likely 159A, who would be considered the primary burial from the tomb. Individuals 159B–E would be formerly interred individuals who had each been pushed aside to make room for each subsequent new interment.

Preservation

The preservation of an individual skeleton, its completeness, degree of fragmentation and the integrity of the bones and teeth, will determine how complete and detailed an osteo-biography can be: the better the preservation, the better the osteobiography. In terms of the preservation of the individuals from Tomb 159, preservation could be rated as poor (159A, B, Δ) or very poor (159Γ, E).

Skeleton 159A: was 25–50% complete and was missing portions of the cranium, including the mandible, the right ulna, most of the left hip, the left fibula, most of the meta-tarsals and pedal phalanges and most of the spine and ribs. None of the long bones was complete with most missing their epiphyses. Most bones were fragmented and exhibited erosion of cortical bone surfaces. Most of the bones had been stained white, a consequence of lying on the tomb's limestone floor.

Skeleton 159B: was also 25–50% complete. It was missing most of its cranium, ribs, spine, left hip, right femur, right tibia and pedal and manual phalanges. None of the long bones was complete, though some of the epiphyses did survive. Most of the bones were fragmented with erosion of the cortical surfaces. Again, the bones were stained white due to prolonged contact with the limestone of the tomb floor.

Skeleton 159Γ: was 25–50% complete with severe erosion of bone cortical surfaces. In contrast to skeletons 159 A and B, this individual had better preservation of the spine and ribs but was missing most of the bones of arms, legs and cranium. This individual's bones also had less white staining, perhaps indicating less contact with the limestone tomb floor.

Skeleton 159Δ: was slightly more complete than the others (50–75%), based on the presence of more ribs and vertebrae, though this individual was missing the right forearm and most of the right hand. All the long bones were fragmented and most were missing their epiphyses. Like individual 159Γ, the bones had less white staining.

Skeleton 159E: comprised only of a few eroded frontal bone fragments and thus had only a completeness of 0–25%.

Age estimation

As adult ageing methods are inherently biased, best practice is to place individuals into age categories rather than assigning them precise ages (O'Connell 2004). Accordingly, the Bradford ageing categories were used (Young Adult: 18–25 years, Young Middle Adult: 26–35, Old Middle Adult: 36–45, Mature Adult: 46+), which are a slight modification of Buikstra and Ubelaker's (1994) ageing categories. Although the preservation of Skeleton 159A was poor it could have its age assessed due to preservation of the auricular surface of the right hip (Lovejoy *et al.* 1985), the maxillary and mandibular molars (Brothwell 1981) and the proximal epiphysis of the left clavicle (Scheuer and Black 2004). The degenerative changes at the auricular surface provided an age of 30–34 years, the wear pattern on the molars indicated an age of 25–35 and the fusion of the clavicle indicated an age of 30+ years. Individual 159A was placed in the Young Middle Adult (26–35) age category.

The only ageing markers available for assessment on Skeleton 159B were the right auricular surface (Lovejoy *et al.* 1985) and several maxillary and mandibular molars (Brothwell 1981). The degeneration of the auricular surface indicated an age of 30–34 years, while the wear on the molars indicated an age of 25–35. Accordingly, Skeleton 159B was assigned an age category of Young Middle Adult (26–35).

Skeleton 159Γ was assigned an age based on degeneration of the auricular surface (Lovejoy *et al.* 1985), the wear on the maxillary and mandibular molars and the degree of fusion apparent in the sacral vertebrae and sternum (Scheuer and Black 2004). The auricular surface indicated an age of 30–34 years, the molars an age of 25–35, the sacrum, whose first and second vertebrae had just fused, indicated an age of ca. 30 and the unfused appearance of the distal sternum indicated an age of less than 40. Based on these age indicators, skeleton 159Γ was assigned to the age category Young Middle Adult (26–35).

Skeleton 159Δ was aged according to degeneration of the right pubic symphysis (Todd 1921a; 1921b; Brooks and Suchey 1990), wear on the maxillary and mandibular molars (Brothwell 1981) and fusion of the sacral vertebrae (Scheuer and Black 2004). Using the Todd method, the pubic symphysis provided an age of 40–45 years, while the Brooks and Suchey method, provided an age of 35.2±9.4. Sacral fusion indicated at age of 30+, however dental attrition (Brothwell 1981) indicated an age of 17–25. As dental wear can be variable and there was congruence between the ages provided by the pubic symphysis and sacrum, the dental age was given less weight and the individual was assigned to the age category of Old Middle Adult (36–45).

Lastly, Skeleton 159E was difficult to age with precision due to the lack of bone elements present. The six cranial fragments did not present with any standard age markers but, due to their size and fact that they appeared to be comprised of newly developed woven bone, it was determined that the fragments were foetal in nature (Scheuer and Black 2004). Accordingly, this individual was assigned to the non-adult age category of foetus.

Biological sex estimation

Biological sex is usually estimated based on the presence of morphological characteristics in the hip bones (Phenice 1969; Buikstra and Ubelaker 1994) and cranium (Buikstra and Ubelaker 1994) and occasionally on metric traits of the clavicle (Jit and Singh 1966), scapula (Iordanidis 1961), humerus (Stewart 1979), radius (Singh *et al.* 1974) and femur (Pearson and Bell 1917/1919; Stewart 1979). Only two diagnostic traits of the pelvis, the greater sciatic notch and preauricular sulcus, were present in Skeleton 159A and both were scored as male. In the cranium, only the nuchal crest, mastoid process and posterior zygomatic arch were available for assessment and all scored as male. Accordingly, based on pelvic and cranial traits, Skeleton 159A was estimated to be male.

Skeleton 159B only had three pelvic traits with which to estimate biological sex: the greater sciatic notch, preauricular sulcus and acetabulum. The latter two were scored as male but the greater sciatic notch was scored as only possible male. Due to the overall lack of sexually diagnostic traits in this skeleton and the slight ambiguity among those present, it was determined it was more prudent to classify Skeleton 159B as a possible male.

The biological sex of Skeleton 159Γ was estimated based on characteristics observed in the sacrum (overall sacral morphology: scored as possible male) and traits in the cranium: supraorbital margin (possible male), supraorbital ridge (possible female), frontal bossing (female), mental eminence (possible male), gonial angle (possible female) and gonial flaring (possible female). In addition, the estimated height of the individual (see below) was also taken into consideration. Cranial traits are mainly correlated with sex differences associated with robusticity (males are more robust) while differences in pelvic traits are largely due to the obstetric demands placed on the female pelvis (Mays and Cox 2000). As such, when estimating biological sex, more weight is generally placed on pelvic traits as they are more accurate (Byers 2017). This individual had a mix of male and female traits but the one pelvic trait was more masculine. Based on these factors (and height), Skeleton 159Γ was tentatively assigned a biological sex of possible male.

Skeleton 159Δ had many pelvic traits (ventral arc, subpubic concavity, subpubic angle, acetabulum, sacral segments and morphology), cranial traits (supraorbital margin, supraorbital ridge, glabella, frontal bossing, mental eminence, gonial angle, palate shape) and one metric trait (femoral head diameter), with which to estimate biological sex. All but the sacral segments (scored as indeterminate) and sacral morphology (scored as possible male) were scored as male, and all the cranial traits, aside from frontal bossing (female), were scored as male. The femoral head diameter fell within the range of possible male. Based on an overwhelming number of male traits, Skeleton 159Δ's biological sex was estimated to be male.

Because the traits that help to differentiate between biological males and females are largely the result of hormone related changes that occur in puberty, it is difficult to determine the biological sex of non-adults (Scheuer and Black 2000). As Skeleton 159E was a foetus and only comprised a few cranial fragments, it was not possible to provide a biological sex estimate.

Population variability

Due to preservation issues it was only possible to estimate the stature of one individual from Tomb 159. Using Trotter's (1970) methodology, which reconstructs stature based on the measurement of the long bones of the legs and arms, skeleton 159Γ's height was estimated to be 167.08 ±4.32 cm (approximately 5ft 5¾ in), based on the length of the left radius. This individual's height fits within the range of average heights for Armenoi males examined by the author to date (n=27; avg. 165.64 cm; females n=10; avg. 156.78 cm).

The skeletons from Tomb 159 were all examined for the presence of standard skeletal cranial and post-cranial non-metric traits, as advocated by Buikstra and Ubelaker (1994). These traits usually take the form of small bones or ossicles found within cranial sutures, proliferative ossifications, ossification failures resulting in defects or variation in foramen number and location. Skeletal non-metric data may provide insights into gene flow, genetic drift, levels of interbreeding, the processes of assorting mating and populations dissimilarity (Tyrell 2000), however, age/sex associations, inter-trait correlations and population-specific asymmetry must be considered when interpreting the data (Buikstra and Ubelaker 1994). Due to the fragmentary nature or absence of the cranium for skeletons in Tomb 159, most cranial non-metric traits were unobservable. A few traits were recorded for skeletons 159A (ossicle in lambdoid sutures, ossicle in occipito-mastoid sutures, condylar canals, and left divided hypoglossal canal), 159Γ (left supraorbital notch) and 159Δ (supraorbital notches). Similarly, due to the fragmentary nature of most of the post-cranial bones from the Tomb 159 inhabitants, most post-cranial non-metric traits were unobservable. One trait was recorded in Skeleton 159A (accessory transverse foramen) and two traits were observed in Skeleton 159Δ (right Poirier's facet, left double anterior calcaneal facet). No patterns among the non-metric traits emerged.

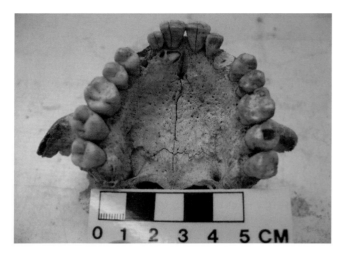

Figure 3.1. Skeleton 159A maxillae with dental caries and impacted right canine.

Dental health and disease

All the adult individuals (159A–Δ) from Tomb 159 had teeth that could be assessed for dental pathology. Skeleton 159A had one caries located on the occlusal surface of the left maxillary second molar (Fig. 3.1). In simple terms, dental caries result from the interaction between oral bacteria (biofilm) and fermentable dietary carbohydrates on a tooth's surface over time (Waldron 2009). This individual was likely eating or drinking foods containing cariogenic sugars in the form of monosaccharides (*e.g.*, fruit, honey) or disaccharides (*e.g.*, fruit, vegetables, milk, beer) or cariogenic carbohydrates such as polysaccharides (*e.g.*, grains, legumes) (Touger-Decker and van Loveren 2003).

Individual 159A also had retention of the maxillary deciduous right canine. The permanent right canine was visible in the bony palate, situated behind the right maxillary central incisor, and had been unable to erupt due to impaction and mesial twisting (Fig. 3.1). Both maxillary canines (the deciduous right and permanent left) had been broken at the root post-mortem. In modern clinical practice, the impaction of permanent maxillary canines is not uncommon, with these teeth being the second most commonly-impacted teeth behind the permanent third molars, with a frequency of 0.8–5.2%, depending on the population (Litsas and Acar 2011). This individual also had variable wear evident in the left and right maxillary molars, with more wear seen on the left side. This pattern was not evident in the mandibular molars, though comparison was difficult, as none of the mandibular teeth was found *in situ* as most of the mandible had been destroyed.

Skeleton 159B had very poor preservation of the dentition. All these teeth were in very poor condition, with most of the teeth having disintegrated, leaving only fragments of root and crown remaining. Both the maxillae and mandible had been destroyed so none of the five teeth present for analysis was *in situ*.

Skeleton 159Γ had a complete mandible but the maxillae had been destroyed, so none of the seven maxillary teeth present was found *in situ*. This individual had considerable wear facets on the labial surfaces of the mandibular anterior teeth, namely the canines and incisors. This wear pattern is suggestive of a pronounced overbite, where the lingual aspects of the maxillary incisors and canines make repeated contact with the labial surfaces of the anterior mandibular dentition (Oltramari-Navarro *et al.* 2010). Interestingly, a similar phenomenon was observed on the teeth of Skeleton 159Δ. This individual had a fragmentary mandible and maxillae. The mandibular teeth were *in situ*, but the maxillary teeth were not. Like skeleton 159Γ, this individual had wear on the labial surfaces of the mandibular anterior teeth, specifically the right canine, lateral and central incisors and the left lateral incisor and canine. Extremely heavy wear was observed on the lingual surfaces of the maxillary canines, matching the wear on the teeth below. However, this wear could not be observed in the maxillary incisors as they had been lost post-mortem. Again, this wear pattern is suggestive of a pronounced overbite (Oltramari-Navarro *et al.* 2010).

Palaeopathology

Only one individual in Tomb 159 (Skeleton 159Δ) exhibited any evidence for skeletal pathology. It is not unusual for individuals to show no sign of pathology on their bones (Weston 2020). It is estimated that, in archaeological contexts, only 15% of the skeletons recovered will have evidence for pathological lesions. This phenomenon may be due to individuals dying from acute conditions that only affect the soft tissues and not the bone or because the individual's immune system did not have enough time to effect change to the skeleton before death occurred (Ortner 2003). Some have argued that those who died without signs of skeletal pathology may in fact be unhealthier than those with chronic skeletal lesions. This argument, called the 'osteological paradox', proposes that those individuals with bone lesions had stronger immune systems that were able to initiate and sustain a lasting response to the disease in question (Wood *et al.* 1992).

Skeleton 159Δ had ossification of the ligamenta flava in several thoracic and lumbar vertebrae. Ligamenta flava are a series of ligaments that connect the laminae of adjacent vertebrae and assist the spinal column in the maintenance of upright posture (Cramer 2014). Ossification of these ligaments is very common and is found in all populations (Mann and Murphy 1990). It may simply be a consequence of ageing (Waldron 2009). This individual also had a very well-healed Colles' fracture at the distal end of the left radius (Fig. 3.2). Colles' fractures result when static and dynamic forces work in opposition, typically in the course

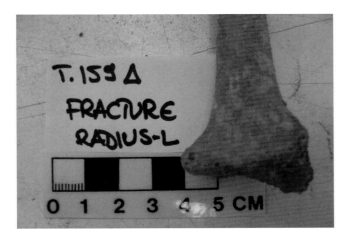

Figure 3.2. Skeleton 159Δ left radius with healed Colles' fracture.

of an accidental fall on an outstretched hand (Grauer and Roberts 1996). In clinical practice, fractures such as this are most commonly seen in older women, who suffer falls associated with ageing and osteoporosis (Summers and Fowles 2022).

Osteobiography summaries

Skeleton 159A belonged to a male individual aged 26–35. He had one dental caries in his upper left second molar and retention of his deciduous upper right canine. There was no evidence for skeletal pathology and the cause of death is unknown.

Skeleton 159B belonged to a possible male individual aged 26–35. There was no evidence for skeletal pathology and the cause of death is unknown.

Skeleton 159Γ belonged to a possible male individual aged 26–35. Stature for this individual was ca. 167 cm. He likely had a pronounced overbite. There was no evidence for skeletal pathology and the cause of death is unknown.

Skeleton 159Δ belonged to a male individual aged 36–45. He likely had a pronounced overbite. There was ossification of the ligamenta flava in their spine and a well-healed Colles' fracture was apparent in their left radius. The cause of death is unknown.

Skeleton 159E belonged to a foetus. The cause of death is unknown but could have been caused by congenital and/or genetic abnormalities, diabetes, hypertension, infection, intrapartum complications or placental dysfunction (Maslovich and Burke 2022).

Discussion and conclusions

Analysis of the individuals buried in Tomb 159, the largest and most prestigious tomb in the Necropolis, has

provided greater insight into the lives of these Late Minoans. There were five individuals interred in the tomb. This number appears to have little significance as the MNI for tombs across Armenoi ranges from one to 28, with no discernible pattern.

The preservation of the skeletons of the individuals contained within the tomb was poor, with a great deal of fragmentation and erosion of the cortical bone. It appears that limestone tombs do not necessarily contribute to long term preservation of the body so, if this was a societal aim, it was a failure. However, community members would have been well aware of the degree of skeletal preservation that was occurring within the tombs as tombs were frequently re-opened for additional interments and previous interments were handled and moved aside. It was likely that, when shifting the remains to the sides of the tomb, the bones on the previous interred individuals were damaged, fragmented, and abraded.

Aside from the foetus (159E), whose biological sex could not be determined, this tomb appears to be a burial place for males, or was it? On the Greek Mainland during the Late Helladic III period, there was a clear segregation of male and female roles, in addition to differentiation based on status and age (Mee 1998).

Segregation of the sexes was not apparent at the Late Minoan III Necropolis however. As reported in Chapter 5 below, 23 people were analysed, using in depth Next-generation sequencing (NGS) and sexed genetically. The result was 16 females and seven males. The ratio, based on previous osteoarchaeological work on Minoan cemeteries, was 1:1 male on a small sampling in 1983, hence it can only be used as a general guideline, that males and females were not differentiated on the basis of sex on their rights to burial in the Late Minoan III Necropolis of Armenoi, and that male and female were treated equally in death. It is therefore almost certain that, in Tomb 159, high status trumped gender identity.

Analysis of Late Minoan II–IIIB burials across Crete has suggested that a mainland pattern was mirrored, as certain funerary objects appear to be associated with particular representations of identity (D'Agata 2020). For example, males appeared to have a burial 'kit' that included items such as weaponry, sharp bronze tools, bronze vessels, metal or clay kylikes, kraters and conical rhyta while female burial 'kits' were less prescribed, though jewellery seems to have been a key component (D'Agata 2020). As Tomb 159 was robbed, it cannot be determined if any of the funerary objects contained within match the gender identities of the Tomb 159 individuals.

Kylikes

Although the presence of kylikes may fit in general with a biological male identity, this cannot be said about all Minoans. The superb Camp Stool frescoes which depict a 'toasting' ceremony is testament to this (Evans 1935, pl. XXXI A–H). Hence this cannot be said about the presence of kylikes in the chambers of tombs at Armenoi. In addition, the greatest number of kylikes were found in the dromoi, and who participated in the ritual ceremonies that took place there, male, female or both, is unknown (Martlew and Kolivaki in prep.)

Jewellery

In the case of Tomb 159, the tomb was robbed and any important jewellery, any gold, would have been the primary target of tomb robbers. The single piece of gold that was found is a leaf bead/pendant (R.M.M. 3241, see Chapter 9). A known example of the burial of a high status female in Late Minoan III is in Tholos A at Archanes (Sakellarikis and Sakellarikis 1997). The fact remains that we have very little idea what adornments high status males, such as a Minoan king or ruler, would wear (but see Chapter 9 and Chapter 10, postscript).

Age of death

In terms of age at death, three of the four adult males died between the ages of 26 and 35, with one who had died between the ages of 36 and 45 years. Angel (1947) proposed a mean age at death for individuals from Middle Bronze Age Greece at 34.7 years and from the Late Helladic III period at 34.1 years, though admitted that his dataset was too small to draw any grand conclusions. Based on her analysis of 250 individuals from Armenoi, McGeorge (1983) estimated mean age at death for males to be 30.67 years and 27.56 years for females. Accordingly, the ages of death for the adult males in Tomb 159 appear to fall within these ranges.

Stature

Stature could only be estimated for one individual from the tomb, 159Γ. The estimated height of 167.08±4.32 cm comes close to the mean height for Armenoi males (167.6 cm) reported by McGeorge (1983). These estimates are not far from the average male heights reported for modern Cretans: 168.1 cm (Poulianos 1971) or 168.6 cm (Hawe 1910). That individual 159Γ had essentially attained the height of the average modern Cretan male indicates that he had likely achieved his genetic potential, height-wise and, accordingly, had not suffered from conditions or malnutrition in childhood that may have affected growth. This could have been related to his status, as economically advantaged adults have a greater propensity to have increased net nutrition, the difference between nutritional intake and nutritional losses due to disease (Deaton 2007).

Dental health

The dental health of the individuals from Tomb 159 was generally good, though one individual (159A) had dental caries. As previously mentioned, dental caries is a result of high carbohydrate diet, so this individual was likely merely consuming a typical Minoan cuisine, which featured carbohydrate rich foods such as grapes, figs, honey, pear, quince, dates, wheat, barley, pulses and pomegranates (Hood 1971; Tsafou and Garcia-Granero 2021). Caries is typically a predictor of lower social status in modern dental practice (Schwendicke *et al.* 2015) but not in antiquity, where individual lifestyle and dental hygiene, rather than socioeconomic status, has a greater influence on caries frequencies (Stranska *et al.* 2015).

Health and disease

In terms of skeletal health and disease, aside from a well-healed wrist fracture and ossified spinal ligaments (Skeleton 159Δ), there was no pathology evidence on the bones of the Tomb 159 individuals. As previously mentioned, this could mean either that they were relatively healthy prior to death and died of acute causes that left no trace (Ortner 2003) or that they were victims of the 'osteological paradox' and were in fact unhealthy individuals whose immune systems were functioning so poorly that they were unable to mount a bony response to their illnesses before their demise (Wood *et al.* 1992). Unfortunately, it is not possible to determine which scenario was correct, however, a lack of joint pathology, as well as a lack of enthesopathies, may be able to say something about the performance of strenuous tasks and manual labour, which may be related to social status.

The well-healed wrist, or Colles', fracture evident in individual 159Δ, signifies that this person likely had access to medical care. It is believed that the Minoans likely had physicians attached to the palace complexes, while communities had priest-healers practising community or folk-medicine (Arnott 1996). There is evidence for high status individuals, like a woman buried in Grave Circle B at Mycenae, having sophisticated medical treatment with the healing of a complex compound fracture of the right humerus (Angel 1973). McGeorge (1983) has reported on other well-healed fractures among individuals at the Necropolis. Accordingly, a well-healed fracture may have either been a high status sign of access to a physician or a non-indicator of status if the individual was treated by a local community practitioner.

Enthesopathy, or activity related pathology, can appear as lytic or proliferative lesions at the sites where tendons and ligaments attach to bones or joint capsules. Their presence can be interpreted as a sign of increased muscle or ligament activity; however, they cannot be correlated to specific actions or occupations – only movements can be inferred (Knüsel 2000). As there were no signs for enthesopathies among the Tomb 159 individuals, it was likely that they were not consistently engaging in strenuous activities, something that might be fitting with high social status. Joint disease can take many forms but the most common type is osteoarthritis (OA), a disease affecting the joint cartilage which breaks down as the disease progresses. The bone of the affected joint responds by producing additional bone to repair the damage and stabilise the joint. Movement, together with age (it is uncommon for individuals under the age of 40 to develop OA), are the main precipitating factors, though genetics, ethnicity and obesity also play a role (Waldron 2009). Considering the ages at death of the Tomb 159 inhabitants (only one may have been over 40), a lack of OA in the spine, large joints of the arms and legs and small joints of the hands and feet, is not unusual but it may also, like a lack of enthesopathies, be a signifier for a lack of consistent, repetitive movement and potential evidence for high social status.

Relationship

Two individuals, 159Γ and 159Δ, had pronounced over-bites that had resulted in marked wear on the mandibular anterior teeth. Pronounced overbites, particularly of the variety observed here, where the lower teeth are essentially covered, abraded and worn by the upper teeth, are under strong genetic influence (Peck *et al.* 1998). Accordingly, its presence in two individuals from the same family tomb is a confirmation of the family ties between individuals buried in Tomb 159.

Summary

Despite poor skeletal preservation, it was possible to construct basic osteobiographies for the inhabitants of Tomb 159, the largest, and most prestigious in the Necropolis. The five tomb occupants, four adult males and one foetus, had some accompanying grave goods that were consistent with their male biological sex and, aside from the foetus, they had all died within the life-expectancy range for individuals of that time period. A genetic relationship between two individuals was indicated by the sharing the same heritable severe over-bite and the presence of caries in one individual signalled consumption of the typical high carbohydrate Minoan diet. The calculated height for one individual provided evidence for attainment of stature typical for the population and was even similar to modern Cretans. No cause of death could be determined for these individuals, but a lack of pathological lesions, particularly those related to joint disease and activity-related pathology, could indicate that the adult individuals buried within Tomb 159 did not engage in strenuous activities, a potential signifier of the high status that was signalled by the tomb's notable architecture features.

Bibliography

Angel, J.L. (1947) The length of life in Ancient Greece. *Journal of Gerontology* 2, 18–24. [https://doi.org/10.1093/geronj/2.1.18]

Angel, J.L. (1973) Human skeletons from Grave Circles at Mycenae. In G.E. Mylonas (ed.), *Grave Circle B of Mycenae*, 379–97. Athens, Archaeological Society of Athens.

Arnott, R. (1996) Healing and medicine in the Aegean Bronze Age. *Journal of the Royal Society of Medicine* 89, 265–70.

Brooks, S.T. and Suchey, J.M. (1990). Skeletal age determination based on the os pubis: a comparison of the Acsádi-Nemeskéri and Suchey-Brooks methods. *Human Evolution* 5, 227–38. [https://doi:10.1007/BF02437238]

Brothwell, D.R. (1981) *Digging Up Bones: the excavation, treatment and study of human skeletal remains*. Cornell NY, Cornell University Press.

Buikstra, J.E. and Ubekaler, D.H. (eds) (1994) *Standards for Data Collection from Human Skeletal Remains*. Fayetteville AR, Arkansas Archaeological Survey.

Byers, S.N. (2017) *Introduction to Forensic Anthropology* (5th edn). London, Routledge.

Cramer, G.D. (2014) The cervical region. In S.A. Darby and G.D. Cramer (eds), *Clinical Anatomy of the Spine, Spinal Cord, and ANS* (3rd edn), 135–209. Amsterdam, Elsevier Health Sciences. [https://doi.org/10.1016/B978-0-323-07954-9.00005-0]

D'Agata, A.L. (2020) Funerary practices, female identities, and the clay pyxis in Late Minoan III Crete. In J.M.A. Murphy (ed.), *Death in Late Bronze Age Greece: variations on a theme*, 300–19. Oxford, Oxford University Press. [https://doi.org/10.1093/oso/ 9780190926069.003.0014]

Deaton, A. (2007) Height, health, and development. *Proceedings of the National Academy of Sciences 104*, 13232–7. [https://doi.org/10.1073/pnas.0611500104]

Evans, A.J. (1935) *The Palace of Minos at Knossos* 4(2). London, Macmillan

Grauer, A. and Roberts, C (1996). Paleoepidemiology, healing, and possible treatment of trauma in the medieval cemetery population of St. Helen-on-the-Walls, York, England. *American Journal of Physical Anthropology* 100, 531–44. [https://doi.org/10.1002/(SICI)1096-8644(199608)100:4<531::AID-AJPA7>3.0.CO;2-T]

Hawes, C.H. (1910) Archaeological and ethnological researches in Crete. Interim report. *Report of the 79th Meeting of the British Association for the Advance of Science, Winnipeg 1909*, 287–91. London, John Murray.

Hood, S. (1971) *The Minoans: the story of Bronze Age Crete*. Westport CO, Praeger.

Iordanidis, P. (1961) Détermination du sexe par les os du squelette (atlas, axis, clavicule, omoplate, sternum). *Annales de Médecine Légale* 41, 280–91.

Jit, I. and Singh, S. (1966). The sexing of adult clavicles. *Indian Journal of Medical Research* 54, 551–71.

Knüsel, C. (2000). Bone adaptation and its relationship to physical activity in the past. In M. Cox, S. Mays (eds), *Human Osteology in Archaeology and Forensic Science*, 381–402. London, Greenwich Medical Media.

Litsas, G. and Acar, A. (2011) A review of early displaced maxillary canines: etiology, diagnosis and interceptive treatment. *Open Dentistry Journal* 5, 39–47. [https://doi.org/10.2174/1874210601105010039]

Lovejoy, C.O., Meindl, R.S., Pryzbeck T.R. and Mensforth, R.P. (1985) Chronological metamorphosis of the auricular surface of the ilium: a new method for the determination of adult skeletal age at death. *American Journal of Physical Anthropology* 68, 15–28. [https://doi: 10.1002/ajpa.1330680103]

Löwe, W. (1996) *Spätbronzezeitliche Bestattungen auf Kreta*. Oxford, British Archaeologial Report S642. [https://doi.org/10.30861/9780860548270]

Mann, R.W. and Murphy, S.P. (1990) *Regional Atlas of Bone Disease*. London, Charles C. Thomas.

Maslovich, M.M. and Burke, L.M. (2022) *Intrauterine Fetal Demise.* StatPearls https://www.ncbi.nlm.nih.gov/books/NBK557533/

Mays, S. and Cox, M. (2000) Sex determination in skeletal remains. In M. Cox and S. Mays (eds), *Human Osteology in Archaeology and Forensic Science*, 117–30. London, Greenwich Medical Media.

McGeorge, P.J.P. (1983) The Minoans: demography, physical variation and affinities. Unpublished PhD thesis, University of London.

Mee, C. (1998) Gender bias in Mycenaean mortuary practices. In K. Branigan (ed.), *Cemetery and Society in the Aegean Bronze Age*, 165–70. Sheffield, Sheffield Academic Press.

O'Connell, L. (2004) Guidance on recording age at death in adults. In M. Brickley and J. McKinley (eds), *Guidelines to the Standards for Recording of Human Remains*, 18–20. Reading, Institute for Field Archaeologists Paper 7/BABAO.

Oltramari-Navarro, P.V.P., Janson, G., Salles de Oliveira, R.B., Quaglio, C.L, Henriques, J.F.C., de Carvalho Sales-Peres, S.H. and McNamara, J.A. (2010) Tooth-wear patterns in adolescents with normal occlusion and Class II Division 2 malocclusion. *American Journal of Orthodontics and Dentofacial Orthopedics* 137. [https://doi.org/10.1016/j.ajodo.2010.01.020]

Ortner, D.J. (2003) *Identification of Pathological Conditions in Human Skeletal Remains* (2nd edn). London, Academic Press.

Papadopoulou, E. (2017) LM III mortuary practices in West Crete: the cemeteries of Maroulas and Armenoi near Rethymnon. *Studi Micenei ed Egeo-Anatolici Nuova Serie* 3, 131–57.

Pearson, K. and Bell, J. (1917/1919). A study of the long bones of the English skeleton. I. the femur. *Drapers' Company Research Memoirs. Biometric Series* X. London, Department of Applied Mathematics, University College London.

Peck, S., Peck, L. and Kataja, M. (1998) Class II Division 2 malocclusion: a heritable pattern of small teeth in well-developed jaws. *Angle Orthodontist* 68, 9–20. [https://doi.org/10.1043/0003-3219(1998)068<0009:CIDMAH>2.3.CO;2]

Phenice, T. (1969) A newly developed visual method of sexing in the os pubis. *American Journal of Physical Anthropology* 30, 297–301. [https://doi.org/10.1002/ajpa.1330300214]

Poulianos, A.N. (1971) *The Origin of the Cretans: anthropological investigations on the Island of Bravery.* Athens, Kostas Koulouthakos Press.

Sakellarakis, I. and Sakellarakis, E. (1997) *Archanes: Minoan Crete in a new light.* Athens, Ammos.

Scheuer, L. and Black, S. (2000) *Developmental Juvenile Osteology.* London, Academic Press.

Scheuer, L. and Black, S. (2004) *The Juvenile Skeleton.* Amsterdam, Elsevier.

Schwendicke, F., Dorfer, C.E., Schlattmann, P., Foster Page, L., Thomson, W.M. and Paris, S. (2015). Socioeconomic inequality and caries: a systematic review and meta-analysis. *Journal of Dental Research* 94, 10–18. [https://doi.org/10.1177/0022034514557546]

Singh, G., Singh, S.P. and Singh, S. (1974) Identification of sex from the radius. *Journal of the Indian Academy of Forensic Sciences* 13, 10–16.

Stewart, T.D. (1979) *Essentials of Forensic Anthropology.* London, Charles C Thomas.

Stranska, P., Veleminsky, P. and Polacek, L. (2015). The prevalence and distribution of dental caries in four early medieval non-adult populations of different socioeconomic status from Central Europe. *Archives of Oral Biology* 60, 62–76. [https://doi.org/10.1016/j.archoralbio.2014.08.002]

Summers, K. and Fowles, S.M. (2022) *Colles fracture.* StatPearls https://www.ncbi.nlm.nih.gov/books/NBK553071

Todd, T.W. (1921a) Age changes in the pubic bone. I: the male white pubis. *American Journal of Physical Anthropology* 3, 285–334. [https://doi.org/10.1002/ajpa.1330030301]

Todd, T.W. (1921b) Age changes in the pubic bone. III: the pubis of the white female. IV: the pubis of the female white-negro hybrid. *American Journal of Physical Anthropology* 4, 1–70. [https://doi.org/10.1002/ajpa.1330040102]

Touger-Decker, R. and van Loveren, C. (2003) Sugars and dental caries. *American Journal of Clinical Nutrition* 78, 881–92S. [https://doi.org/10.1093/ajcn/78.4.881S]

Trotter, M. (1970) Estimation of stature from intact limb bones. In T.D. Stewart (ed.), *Personal Identification in Mass Disasters*, 71–84. Washington DC, National Museum of Natural History.

Tsafou, E. and Garcia-Granero, J.J. (2021) Beyond staple crops: exploring the use of 'invisible' plant ingredients in Minoan cuisine through starch grain analysis on ceramic vessels. *Archaeological and Anthropological Sciences* 13, 128. [https://doi.org/10.1007/s12520-021-01375-4]

Tyrell, A. (2000) Skeletal non-metric traits and the assessment of inter- and intra-population diversity: past problems and future potential. In M. Cox and S. Mays (eds), *Human Osteology in Archaeology and Forensic Science,* 289–306. London, Greenwich Medical Media.

Tzedakis, Y. and Kolivaki, V. (2018) Background and history of the excavation. In Y. Tzedakis, H. Martlew and R. Arnott (eds), *The Late Minoan III Necropolis of Armenoi 1.,* 1–18. Philadelphia PA, INSTAP Academic Press. [available at: https://doi.org/10.2307/j.ctvggx2sj.9]

Tzedakis, Y. and Martlew, H. (eds) (1999) *Minoans and Mycenaeans Flavours of their Time.* Athens, Greek Ministry of Culture, General Directorate of Antiquities.

Tzedakis, Y. and Martlew, H. (2007) A chorotaxia at the Late Minoan III Cemetery of Armenoi. In P.P. Betancourt, M.C. Nelson and H. Williams (eds), *Krinoi Kai Limenes: Studies in Honor of Joseph and Maria Shaw*, 67–73. Philadelphia PA, INSTAP Academic Press.

Waldron, T. (2009) *Palaeopathology.* Cambridge, Cambridge University Press.

Weston, D.A. (2020). Human osteology. In: M. Richards and K. Britton (eds), *Archaeological Science: an introduction,* 147–69. Cambridge, Cambridge University Press. [https://doi:10.1017/9781139013826.007]

Wood, J., Milner, G., Harpending, H. and Weiss, K. (1992) The osteological paradox: problems of inferring skeletal health from prehistoric samples. *Current Anthropology* 33, 343–70.

4

Multi-isotopic (C, N, S, Sr) measurements of human skeletal material

Michael P. Richards

Introduction

Previous isotope dietary studies using carbon and nitrogen isotope values of humans from Armenoi showed that most humans from the site had a similar diet that was high in animal protein with little, or no, contribution from marine proteins (Richards and Hedges 2008). These previous studies showed no measurable difference in dietary isotope values between males and females. Also, perhaps surprisingly, there were no isotope dietary differences between individuals buried in more elaborate tombs with grave goods and those buried in simpler and smaller tombs without grave goods. The Armenoi Necropolis spans the Late Minoan III period (Tzedakis *et al.* 2018), following LM II when Mycenaean Linear B text first appears on Crete (Palaima 2010). However, there were no clear changes in carbon and nitrogen dietary isotope values over these time periods (Richards and Hedges 2008).

To expand the isotope studies of the Armenoi humans, additional isotope systems, specifically sulphur and strontium, were measured on a number of individuals from the site. These isotopes are more associated with migration and mobility studies and have the potential to identify geographical 'outliers' in a burial population. Of particular interest with these isotopes is to look for potential 'outliers' at Armenoi that either originated from elsewhere on Crete or, indeed, if any of the individuals buried at Armenoi, especially in the late periods, may have originated from outside Crete, including the Mycenaean controlled Greek mainland.

The results of the strontium and sulphur isotope measurements of the Armenoi humans are presented below. Also included are updated carbon and nitrogen isotope values using newly prepared samples.

Isotope analysis background

Isotope analysis of human bone and teeth is an established method for determining past diets and mobility (Britton 2017; Richards 2020). Carbon and nitrogen isotopes of bone collagen reflect the sources of dietary protein over a number of years of life and are often used to determine the relative amounts of terrestrial foods compared with marine foods in diets, the consumption of C4 plant foods (*e.g.* maize or millet) and the relative amounts of animal compared to plant proteins in diets. In the Armenoi context specifically, the method is useful for determining the relative amounts of animal (*i.e.* meat, milk) and plant (*i.e.* cereals) proteins in diets (Richards 2015).

Sulphur isotopes are also measured in bone collagen and are an indicator of both diets and mobility (Nehlich 2015). They are used to trace the source of dietary proteins in human diets, especially if dietary proteins were mainly from local sources or from other locations or ecosystems which have different sulphur isotope values than the local values. One particular use of them is to determine if freshwater foods were consumed regularly (Nehlich *et al.* 2010). This is because freshwater ecosystems often have very different sulphur isotope values than terrestrial ecosystems. Similarly, marine foods often have different sulphur isotope values than terrestrial foods, so they can also be used to assess the consumption of marine foods. It can also be used as a mobility indicator if some individuals have sulphur values that differ greatly from the sulphur values of local foods and, especially, if the values are very different from other humans buried at the same site.

Strontium isotopes are mainly used as a mobility indicator (Britton 2020). They are measured in tooth enamel and the strontium ratios relate to the geographical location where the individual was living when that tooth was formed. The third molar or second premolar is the most often used tooth for adults for strontium isotope studies and the enamel strontium records where the individual was living in later childhood.

As sulphur isotopes are measured in bone collagen they can provide a measure of location in the latter parts of life, whereas strontium in tooth enamel records childhood location. Combining these two isotope systems can then better help determine an individual's geographical history. Those with non-local strontium isotope values but local sulphur values may have then lived elsewhere as children but then spent the majority of their life in the local area. Those that have non-local strontium and sulphur were likely non-local in childhood and adulthood, despite then being buried in the local site.

For both strontium and sulphur isotope values it is necessary to establish the local, baseline, strontium and sulphur values (Bentley 2006; Nehlich 2015). Strontium baseline values are often obtained using modern plant samples from the local region (Wong *et al.* 2021). Where this is not possible, strontium values of archaeological animal remains, especially those species that were likely raised locally, are measured (Bentley 2006). Sulphur baseline values are also usually obtained using local archaeological faunal samples (Nehlich 2015).

Sample selection

The first isotope studies at Armenoi were for dietary studies using carbon and nitrogen isotope values and samples were selected in the late 1990s by H. Martlew and M. Richards (Tzedakis and Martlew 1999; Richards and Hedges 2008). As this method uses bone, teeth were not sampled at this stage. Later sampling of a number of individuals was undertaken by H. Martlew, and this included a number of tooth samples that were subsequently measured for carbon, nitrogen, strontium and sulphur. In addition, tooth samples were taken specifically for strontium and sulphur isotope measurements by M. Richards in 2007. The complete list of all the sampled humans (total of 84 bone samples and nine enamel samples) and their isotope values is given in Table 4.1. Additional samples were subsequently analysed for isotopic data as part of the programme of ancient DNA analysis. These are reported on in Chapter 5.

Methodology

Sample preparation and measurement was undertaken at the (former) Department of Human Evolution at the Max Planck Institute for Evolutionary Anthropology in Leipzig, Germany. For carbon and nitrogen isotope analysis, all samples were prepared using the procedures outlined in

Richards and Hedges (1999) and Richards (*et al.* 2022) and which includes the use of ultrafiltration of the extracted collagen (Brown *et al.* 1988). These samples include those that were originally prepared in 1997/1998 and previously published (Richards and Hedges 2008). These earlier measurements were made on collagen that was not ultrafiltered, so for this study we reprepared (from bone and teeth) collagen and undertook new isotope measurements. Also included are new carbon and nitrogen measurements on samples that were not previously published or measured. Sulphur isotope measurements of ultrafiltered collagen were undertaken following procedures outlined in Nehlich and Richards (2009). Errors on the isotope measurements are (1σ) $\delta^{13}C\pm 0.1‰$, $\delta^{15}N\pm0.2‰$, and $\delta^{34}S\pm0.5‰$, based on long-term measurements of internal secondary and international standards.

The strontium isotope measurements of the faunal and human teeth were made on solutions of purified and isolated strontium extracted from sampled tooth enamel using the preparation protocol outlined in Richards *et al.* (2022). The strontium isotope ratios were measured using a Thermo Fisher Neptune MC-ICP-MS following instrument parameters and data collection methods outlined therein.

The isotope studies of the Armenoi humans is part of a larger project looking at diet and mobility at Neolithic and Bronze Age sites across Crete (Richards *et al.* 2022). This project measured faunal strontium and sulphur isotope values from many sites in Crete and was used to produce local baseline values for different regions of the island and for all of Crete itself. These values are included here to show the average baseline strontium and sulphur values obtained for all of Crete.

Results

The repreparation and remeasurements of the carbon and nitrogen isotope values of the Armenoi humans that were studied previously resulted in values that were somewhat different than the previously published values (Table 4.2). These differences were not in the $\delta^{13}C$ values (average difference of 0.003‰) but mainly in the $\delta^{15}N$ values (average difference of 0.9‰). The difference in the $\delta^{15}N$ values may be related to the use of ultrafiltration for the more recent samples which resulted in the extraction of better-preserved collagen. However, it is most likely that the difference is due to the use of different isotope reference standards for the isotope measurements. Despite the relatively small differences in the isotope values, the conclusions about human diets at Armenoi have not changed. The newly measured carbon and nitrogen isotope values show, as previously reported (Richards and Hedges 2008), human diets high in animal protein, which may have been from meat or dairy, and no measurable input of marine foods in diets.

Table 4.1. List of the individuals sampled for isotope analysis from Armenoi and associated carbon, nitrogen, sulphur and strontium isotope values.

S-EVA	Sample no.	Tomb	Element	$\delta^{13}C$	$\delta^{15}N$	C:N	$\delta^{34}S$	%S	$^{87}Sr/^{86}Sr$
3426	-ARM 500	86 A	Femur	−20.2	7.8	3.34	13.6	0.21	n/a
3331	-ARM 500	86 A	tooth enamel	n/a	n/a	n/a	n/a	n/a	0.709191
3519	-ARM 501	55 Γ	Femur	−19.4	8.8	3.21	17.6	0.23	n/a
3529	-ARM 501	55 Γ	tooth enamel	n/a	n/a	n/a	n/a	n/a	0.709114
3333	ARM-501(38)	55 Γ	tooth enamel	n/a	n/a	n/a	n/a	n/a	0.709415
3335	ARM-502(39)	118 D	tooth enamel	n/a	n/a	n/a	n/a	n/a	0.709122
3427	-ARM 502	118 D	femur	−19.8	9.0	3.56	n/a	n/a	n/a
3428	-ARM 503	89 I (160)	femur	−19.7	8.2	3.29	13.9	0.16	n/a
3337	ARM-503(40)	89 I (160)	tooth enamel	n/a	n/a	n/a	n/a	n/a	0.709032
3429	-ARM 504	89 B (200)	femur	−19.5	8.3	3.19	19.9	0.17	n/a
3339	ARM-504(41)	89 B (200)	tooth enamel	n/a	n/a	n/a	n/a	n/a	0.709228
1265	-ARM 505	89 K	femur	−19.7	9.2	3.30	n/a	n/a	n/a
3430	-ARM 505	89 K	femur	−20.2	8.8	3.33	16.2	0.2	n/a
3431	-ARM 506	93 A	femur	−19.9	8.8	3.35	n/a	n/a	n/a
3500	ARM 506/588	93 S	femur	−19.6	9.0	3.36	n/a	n/a	n/a
3432	-ARM 507	71 E	femur	−19.7	8.6	3.22	n/a	n/a	n/a
3433	-ARM 508	77 A	femur	−19.6	8.4	3.24	14.1	0.17	n/a
3434	-ARM 509	29 G	femur	−19.6	9.4	3.29	13.9	0.22	n/a
3435	-ARM 510	117 G	femur	−19.4	8.6	3.32	12.7	0.19	n/a
3436	-ARM 511	117 A	femur	−19.8	8.6	3.37	13.7	0.22	n/a
3437	-ARM 512	67 B	femur	−19.8	9.2	3.25	13.1	0.2	n/a
3438	-ARM 513	89 ST	femur	−19.7	8.1	3.23	n/a	n/a	n/a
3439	-ARM 514	71 B	femur	−19.6	8.9	3.27	15.2	0.18	n/a
3440	-ARM 515	92 B	femur	−19.8	8.8	3.30	16.2	0.18	n/a
3441	-ARM 516	79 ST	femur	−19.8	8.6	3.31	15.4	0.19	n/a
3442	-ARM 517		femur	−20.0	7.9	3.48	n/a	n/a	n/a
3443	-ARM 518	10-2	skull	−19.9	8.7	3.50	n/a	n/a	n/a
3444	-ARM 519	73 ST	femur	−19.8	8.8	3.35	n/a	n/a	n/a
3445	-ARM 520	78 ST	femur	−19.6	8.7	3.30	15.7	0.27	n/a
3520	-ARM 521	69 A	femur	−19.6	9.0	3.22	13.0	0.18	n/a
3531	-ARM 521	69 A	tooth enamel	n/a	n/a	n/a	n/a	n/a	0.709226
3521	-ARM 522	89 A	femur	−19.5	8.8	3.24	13.8	0.21	n/a
3533	-ARM 522	89 A	tooth enamel						0.709153
3446	-ARM 523	86 ST	femur	−19.6	8.3	3.26	13.6	0.15	n/a
3522	-ARM 524	69 B	femur	−20.2	8.4	3.33	n/a	n/a	n/a
3524	-ARM 524	69 B	tooth enamel	n/a	n/a	n/a	n/a	n/a	0.708741
3447	-ARM 525	78 A	femur	−19.6	8.7	3.27	14.3	0.21	n/a
3448	-ARM 526	79 Γ	femur	−19.7	8.4	3.33	14.3	0.21	n/a
3449	-ARM 527	86 Γ	femur	−19.7	9.1	3.36	11.2	0.22	n/a
3525	-ARM 528	89 Γ	femur	−19.7	8.5	3.37	n/a	n/a	n/a
3527	-ARM 528	89 Γ	tooth enamel						
3450	-ARM 529	76 D	femur	−19.6	9.1	3.35	17.6	0.21	n/a

(Continued)

Table 4.1. *List of the individuals sampled for isotope analysis from Armenoi and associated carbon, nitrogen, sulphur and strontium isotope values.* (Continued)

S-EVA	Sample no.	Tomb	Element	$\delta^{13}C$	$\delta^{15}N$	C:N	$\delta^{34}S$	%S	$^{87}Sr/^{86}Sr$
3451	-ARM 530	201 Γ	skull	−19.6	9.9	3.29	12.2	0.17	n/a
3452	-ARM 531		femur	−20.1	8.3	3.39	n/a	n/a	n/a
3453	-ARM 532		femur	−19.8	8.9	3.38	10.0	0.21	n/a
3454	-ARM 533		femur	−19.7	8.0	3.21	13.3	0.16	n/a
3501	-ARM 534		femur	−19.8	7.9	3.27	12.8	0.17	n/a
3455	-ARM 535		femur	−19.7	8.4	3.28	13.3	0.19	n/a
3502	-ARM 536	95 L	femur	−19.8	8.5	3.27	17.7	0.21	n/a
3503	-ARM 537	95 D	femur	−19.7	8.7	3.27	13.2	0.18	n/a
3504	-ARM 538	95 K	femur	−19.7	9.1	3.28	14.4	0.21	n/a
3505	-ARM 539	95 Z	femur	−20.6	8.5	3.75	n/a	n/a	n/a
3506	-ARM 540	95 E	femur	−20.1	9.0	3.33	n/a	n/a	n/a
3456	-ARM 541		femur	−20.1	8.2	3.21	14.1	0.24	n/a
3457	-ARM 542		femur	−20.0	8.3	3.17	14.0	0.19	n/a
3458	-ARM 543		femur	−19.6	9.6	3.22	12.4	0.21	n/a
3459	-ARM 544		femur	−19.6	8.8	3.29	12.6	0.14	n/a
3460	-ARM 545		femur	−19.9	8.9	3.24	14.5	0.2	n/a
3461	-ARM 546		femur	−20.3	8.2	3.48	14.0	0.2	n/a
3462	-ARM 547		femur	−19.6	8.6	3.64	n/a	n/a	n/a
3463	-ARM 548		femur	−19.7	8.5	3.21	14.5	0.23	n/a
3464	-ARM 549		femur	−19.8	8.5	3.22	13.5	0.18	n/a
3465	-ARM 550		femur	−20.3	8.4	3.45	12.8	0.16	n/a
3466	-ARM 551		femur	−19.6	8.6	3.31	13.5	0.19	n/a
3507	-ARM 552	95 ST	femur	−19.7	8.3	3.39	12.0	0.2	n/a
3508	-ARM 553	95 D	femur	−20.0	8.5	3.34	n/a	n/a	n/a
3509	-ARM 554	95 H	femur	−19.7	8.2	3.39	n/a	n/a	n/a
3467	-ARM 555		femur	n/a	n/a	n/a	12.3	0.19	n/a
3468	-ARM 556		femur	−20.2	8.5	3.60	12.5	0.15	n/a
3469	-ARM 557		femur	−19.7	9.4	3.40	13.3	0.18	n/a
3470	-ARM 558		femur	−19.9	8.6	3.30	11.8	0.15	n/a
3471	-ARM 559		femur	−19.9	8.6	3.32	n/a	n/a	n/a
3472	-ARM 560		Femur	−19.7	9.6	3.47	n/a	n/a	n/a
3473	-ARM 562		femur	−19.7	8.0	3.26	12.4	0.17	n/a
3474	-ARM 563		femur	−19.8	7.8	3.31	11.5	0.22	n/a
3475	-ARM 564		femur	−19.5	8.3	3.35	12.1	0.14	n/a
3477	-ARM 566		femur	−19.7	8.3	3.30	14.1	0.24	n/a
3478	-ARM 567		femur	−19.8	8.1	3.36	13.9	0.19	n/a
3479	-ARM 568		femur	−19.9	8.3	3.38	n/a	n/a	n/a
3480	-ARM 569		femur	−19.8	8.9	3.28	14.0	0.2	n/a
3483	-ARM 573		femur	−19.7	9.1	3.41	n/a	n/a	n/a
3484	-ARM 574		femur	−19.6	9.0	3.28	13.5	0.22	n/a
3485	-ARM 575		femur	−19.6	9.1	3.29	13.1	0.2	n/a
3487	-ARM 577		femur	−19.6	9.2	3.32	12.2	0.22	n/a

(Continued)

Table 4.1. (Continued)

S-EVA	Sample no.	Tomb	Element	$\delta^{13}C$	$\delta^{15}N$	C:N	$\delta^{34}S$	%S	$^{87}Sr/^{86}Sr$
3510	-ARM 578		femur	−20.0	8.7	3.30	12.3	0.2	n/a
3499	ARM 578/592	55 A	femur	−19.9	8.8	3.22	12.1	0.22	n/a
3488	-ARM 579		femur	−19.6	8.3	3.29	n/a	n/a	n/a
3489	-ARM 580		femur	−19.7	8.4	3.28	15.1	0.26	n/a
3490	-ARM 581		femur	−20.0	8.2	3.30	13.4	0.22	n/a
3491	-ARM 582		femur	−19.8	9.6	3.35	n/a	n/a	n/a
3492	-ARM 583		femur	−19.6	8.8	3.32	n/a	n/a	n/a
3496	-ARM 586	118 B-Z	femur	−19.7	9.1	3.21	10.8	0.19	n/a
3498	-ARM 588	218	femur	−19.9	8.1	3.31	13.5	0.2	n/a

Table 4.2. List of previously published values of Armenoi humans.

Sample	Previous $\delta^{13}C$	New $\delta^{13}C$	Previous $\delta^{15}N$	New $\delta^{15}N$	Previous C:N	New C:N
ARM 500	−19.9	−20.0	6.8	7.8	3.3	3.3
ARM 501	−19.5	−19.8	9.2	8.5	3.2	3.3
ARM 503	−19.8	−19.7	7.4	8.2	3.4	3.3
ARM 504	−19.6	−19.6	7.6	8.6	3.3	3.3
ARM 505	−19.9	−20.2	7.3	8.8	3.0	3.3
ARM 506	−19.7	−19.9	8.2	8.8	3.0	3.4
ARM 506/588	−20.0	−19.6	8.1	9.0	3.4	3.4
ARM 507	−19.8	−19.7	7.5	8.6	3.4	3.2
ARM 508	−19.4	−19.6	7.5	8.4	3.3	3.2
ARM 509	−19.6	−19.6	8.5	9.4	3.3	3.3
ARM 510	−19.2	−19.4	7.8	8.6	3.3	3.3
ARM 511	−19.9	−19.8	8.0	8.6	3.6	3.4
ARM 512	−19.8	−19.8	8.6	9.2	3.3	3.3
ARM 513	−19.6	−19.7	7.3	8.1	3.3	3.2
ARM 514	−19.6	−19.6	7.7	8.9	3.2	3.3
ARM 515	−19.9	−19.8	7.9	8.8	3.3	3.3
ARM 516	−19.7	−19.8	8.2	8.6	3.3	3.3
ARM 518	−19.8	−19.9	7.9	8.7	3.3	3.5
ARM 519	−20.1	−19.8	7.4	8.8	3.5	3.4
ARM 520	−19.7	−19.6	7.7	8.7	3.4	3.3
ARM 521	−19.6	−19.6	8.2	9.0	3.3	3.2
ARM 522	−19.7	−19.5	7.6	8.8	3.3	3.2
ARM 523	−19.5	−19.6	7.4	8.3	3.2	3.3
ARM 524	−19.9	−20.2	8.1	8.4	3.1	3.3
ARM 525	−19.5	−19.6	7.6	8.7	3.3	3.3
ARM 526	−19.7	−19.7	7.1	8.4	3.2	3.3
ARM 527	−19.5	−19.7	8.1	9.1	3.3	3.4
ARM 528	−20.1	−19.7	7.1	8.5	3.5	3.4
ARM 529	−20.0	−19.6	8.0	9.1	3.5	3.3
ARM 530	−19.3	−19.6	8.7	9.9	2.9	3.3

(Continued)

Table 4.2. List of previously published values of Armenoi humans. (Continued)

Sample	Previous δ¹³C	New δ ¹³C	Previous δ ¹⁵N	New δ ¹⁵N	Previous C:N	New C:N
ARM 536	–20.0	–19.8	7.9	8.5	3.3	3.3
ARM 537	–19.8	–19.7	7.8	8.7	3.3	3.3
ARM 538	–19.7	–19.7	8.1	9.1	3.3	3.3
ARM 539	–20.1	–20.6	7.8	8.5	3.4	3.8
ARM 540	–20.1	–20.1	7.3	9.0	3.2	3.3
ARM 552	–19.8	–19.7	7.5	8.3	3.4	3.4
ARM 553	–20.6	–20.0	7.2	8.5	3.3	3.3
ARM 554	–19.8	–19.7	8.1	8.2	3.4	3.4
ARM 578/592	–19.8	–19.9	8.0	8.8	3.3	3.2

Figure 4.1. Armenoi humans: sulphur isotope values plotted with the average faunal sulphur isotope values from Neolithic and Bronze Age sites across Crete (Table 4.3; data from Richards et al. *(2022).*

The sulphur isotope values of the Armenoi humans (Table 4.1) are plotted in Figure 4.1, with the average sulphur value of animal remains from a number of sites across Crete (Richards *et al.* 2022). As can be observed, most of the humans from the site have sulphur isotope values that plot within the range of 'local' sulphur values for contemporary faunal remains from across Crete. This indicates that the foods these individuals were consuming were likely from local sources, or from other parts of Crete. Of note, however, are a number of individuals that have sulphur isotope values that are higher than the local faunal values. These individuals then likely consumed foods from regions outside Crete,

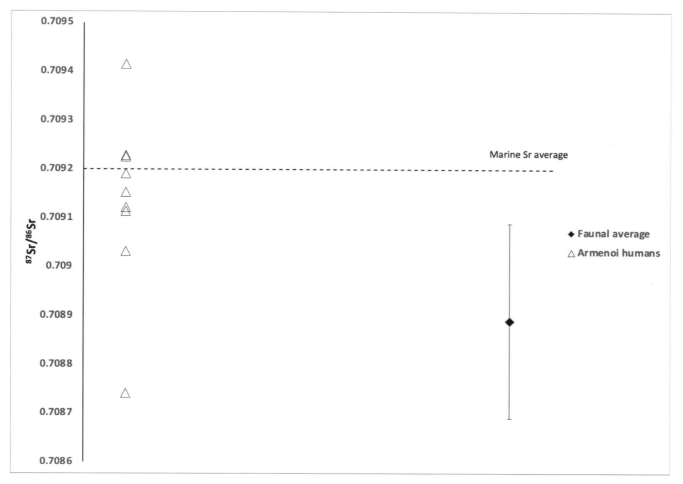

Figure 4.2. Strontium isotope values for Armenoi humans plotted with the average faunal isotope values from Neolithic and Bronze Age sites across Crete.

where the baseline sulphur values are higher than observed across Crete, and therefore are likely immigrants to the Island. Higher sulphur isotope values such as these outliers may indicate a coastal environment, where marine sulphur has been deposited through sea spray or rainfall (Nehlich 2015). As there are currently very few measurements of sulphur isotopes from the eastern Mediterranean region it is not possible to suggest specific geographical regions where these individuals may have arrived from.

The strontium values of the Armenoi humans are plotted in Figure 4.2, along with the average strontium values of faunal remains from other sites in Crete (Richards *et al.* 2022). Similar to the sulphur values, the majority of humans from Armenoi have strontium values that fall within the local range for Crete itself. It is likely then that the majority of these individuals spent their childhoods locally or elsewhere on Crete. There are a number of individuals with higher strontium values than observed for fauna from Crete. And of these three are even higher than the average marine strontium value of 0.7092 (Vizier 1989). These

individuals, especially the three with the highest values, likely did not spend their childhoods (when their tooth enamel was formed) on Crete. Higher strontium isotope values are often found in regions with older bedrock, rather than the limestones that dominate the geology of much of the eastern Mediterranean. As with sulphur, there are few published strontium studies from this region to be used as a comparison.

Two individuals have both strontium and sulphur isotope values that are higher than the faunal averages for Crete and most of the other individuals from Armenoi. These results are given below in Table 4.3. For these two individuals the strontium isotope data indicates that they spent their childhoods outside of Crete, and the sulphur isotope values show that much of their adult lives were also spent away from Crete.

Discussion

The results from the dietary isotope (carbon and nitrogen) studies of the Armenoi humans are perhaps not surprising.

Table 4.3. Sulphur and strontium isotope results from two humans from Armenoi with isotope values outside the range of local faunal values for Crete.

Sample	Sample no.	Tomb	Sex	Element	$^{87}Sr/^{86}Sr$	$\delta^{34}S$
S-EVA 3333	501	55 Г	M	Tooth	0.709415	n/a
S-EVA 3519	501	55 Г	M	Femur	n/a	17.6
S-EVA 3339	504	89 B	M	Tooth	0.709228	n/a
S-EVA 3429	504	89 B	M	Femur	n/a	19.9
Faunal average					0.708890±0.0002 (n=93)	12.2±2.4 (N=35)

They match well with data previously published from the site (Richards and Hedges 2008) and newly prepared samples associated with a large scale DNA study (Chapter 5). The results themselves are also are similar to values from other Bronze Age sites on Crete and from the Mainland (Papathanasiou *et al.* 2015). In most cases it appears that Bronze Age humans from Greece had diets relatively high in animal protein, perhaps mainly from dairy in addition to animal meat. And there is almost no evidence of any measurable marine food consumption, despite many Bronze Age sites being situated on the coast of Crete and mainland Greece. The only place that had any evidence of marine foods was at Mycenae, in the elite burials in Grave Circle A and B (Tzedakis and Martlew 1999; Richards and Hedges 2008). This may point to marine foods being mainly for high status individuals.

The sulphur and strontium values from Armenoi differ from values obtained from other sites in Crete. Most humans from other Cretan sites, both Neolithic and Bronze Age, have sulphur and strontium values that are similar to those measured from contemporary animals from sites across the island. Armenoi though is different. Here, as discussed above, there are a number of individuals with strontium and sulphur isotope values that likely originated outside of Crete. Two individuals in particular, males from tomb 55 Г and 89 B have sulphur and strontium values that indicate their childhood and adult lives were spent away from Crete. Is it possible that these two were new immigrants to Crete at the very end of the Minoan period, associated with the first appearance of Mycenaean Linear B script from the Mainland? Or perhaps were they traders or travellers who originated from outside Crete, who died and were then interred at Armenoi?

Conclusions

The carbon and nitrogen dietary isotope studies of humans from Armenoi have shown that most of the individuals interred in the Necropolis had similar diets, despite the wide range in tomb type and inclusions of grave goods indicated a stratified society. And similar to other Bronze Age sites on Crete and the Mainland, the isotope values indicate diets relatively high in animal protein (meat or dairy) and with no measurable amounts of marine protein (Papathanasiou *et al.* 2015).

The strontium and sulphur studies clearly showed that the majority of individuals buried at Armenoi were from the local region or, perhaps, from other areas of Crete. However, there are some individuals buried at the site that were not local to Armenoi, or even to Crete. These include some individuals that spend their childhoods outside of Crete (base on their enamel strontium values), and some who lived outside of Crete as adults (based on their bone collagen sulphur isotope values). Two individuals lived most of their lives, childhood and adulthood, outside of Crete and, perhaps, these were immigrants from the Mycenaean Mainland, who may have come to Armenoi as traders or settlers. Future DNA research that included these two individuals could tell us more about their potential original geographical origin.

This multi-isotope isotope study of Armenoi has shown the utility of combining dietary and mobility isotope measurements to provide a clearer picture of the lives of those interred at the site. With the expansion of strontium and sulphur isotope measurements to more individuals and especially to other regions to improve our knowledge of the regional baseline variations, and the addition of other isotope methods in the future (including zinc isotopes and compound-specific isotope analysis) it will be possible to further refine our interpretations of diet and mobility at this important Minoan site.

Acknowledgements

Thanks are given especially to Holley Martlew and Yannis Tzedakis for the generosity in providing the opportunity to study the exceptional material from Armenoi. The isotope measurements reported here are part of a larger study and thanks are given to my collaborators Vaughan Grimes, Colin Smith, Olaf Nehlich, Darlene Weston and Keith Dobney.

Thanks is given for assistance with sample preparation and measurements in Leipzig by Annette Weiske, Annabell Reiner and Sven Steinbrenner.

Bibliography

Bentley, R.A. (2006) Strontium isotopes from the earth to the archaeological skeleton: a review. *Journal of Archaeological Method and Theory* 13, 135–87.

Britton, K. (2017) A stable relationship: isotopes and bioarchaeology are in it for the long haul. *Antiquity* 91(358), 853–64.

Britton, K. (2020) Isotope analysis for mobility and climate studies. In M. Richards and K. Britton (eds), *Archaeological Science: an introduction,* 99–124. Cambridge, Cambridge University Press.

Brown, T.A., Nelson, D.E., Vogel, J.S. and Southon, J.R. (1988) Improved collagen extraction by modified Longin method. *Radiocarbon* 30(2), 171–7.

Martlew, H. and Tzedakis, Y. (1999) *Minoans and Mycenaeans: flavours of their time.* Athens, Greek Ministry of Culture.

Nehlich, O. (2015) The application of sulphur isotope analyses in archaeological research: a review. *Earth Science Reviews* 142, 1–17.

Nehlich, O. and Richards, M.P. (2009) Establishing collagen quality criteria for sulphur isotope analysis of archaeological bone collagen. *Archaeological and Anthropological Sciences* 1(1), 59–75.

Nehlich, O., Boric, D., Stefanovic, S. and Richards, M. (2010) Sulphur isotope evidence for freshwater fish consumption: a case study from the Danube Gorges, SE Europe. *Journal of Archaeological Science* 37(5), 1131–9.

Palaima, T.G. (2010) Linear B. In E Cline (ed.), *The Oxford Handbook of the Bronze Age Aegean,* 356–72. Oxford, Oxford University Press.

Papathanasiou, A., Richards, M.P. and Fox, S.C. (2015) *Archaeodiet in the Greek World: Dietary Reconstruction from Stable Isotope Analysis.* Princeton NJ, American School of Classical Studies at Athens.

Richards, M.P. (2015) Stable isotope analysis of bone and teeth as a means for reconstructing past human diets in Greece. In Papathanasiou *et al.* (eds), 15–23.

Richards, M.P. (2020). Isotope analysis for diet studies. In M. Richards and K. Britton (eds), *Archaeological Science: an introduction,* 125–44. Cambridge, Cambridge University Press.

Richards, M. and Hedges, R. (1999) Stable isotope evidence for similarities in the types of marine foods used by late Mesolithic humans at sites along the Atlantic coast of Europe. *Journal of Archaeological Science* 26(6), 717–22.

Richards, M.P. and Hedges, R.E.M. (2008) Stable isotope results from the sites of Gerani, Armenoi and Mycenae. In Y. Tzedakis, H. Martlew and M.K. Jones (eds), *Archaeology Meets Science: biomolecular and site investigations in Bronze Age Greece,* 220–30. Oxford, Oxbow Books.

Richards, M.P., Smith, C., Nehlich, O., Grimes, V., Weston, D., Mittnik, A., Krause, J., Dobney, K., Tzedakis, Y. and Martlew, H. (2022) Finding Mycenaeans in Minoan Crete? Isotope and DNA analysis of human mobility in Bronze Age Crete. *PLoS One* 17(8), e0272144. [https://doi.org/10.1371/journal.pone.0272144]

Tzedakis, Y., Martlew, H. and Arnott, R. (eds) (2018). *The Late Minoan III Necropolis of Armenoi* 1. Philadelphia PA, INSTAP Academic Press.

Viezer, J. (1989) Strontium isotopes in seawater through time. *Annual Reviews of Earth and Planetary Science* 17, 141–67.

Wong, M., Grimes, V., Steskal, M., Song, J., Ng, J., Jaouen, K., Lam, V.C. and Richards. M. (2021) A bioavailable baseline strontium isotope map of southwestern Turkey for mobility studies. *Journal of Archaeological Science Reports* 37, 102922. [https://doi.org/10.1016/j.jasrep.2021.102922]

Bioarchaeological analyses of human and faunal skeletal remains and radiocarbon dating

M. George B. Foody, Peter W. Ditchfield and Ceiridwen J. Edwards

Introduction

Understanding the causes of change seen within the archaeological record is one of the major challenges when interpreting the lives of past societies. This can be addressed directly by applying detailed genetic and stable isotopic analyses to one such rapid period of societal change, that seen after Mycenaean intervention in the Late Bronze Age of Crete. In order to study this key moment in the development of western civilisation, we focused on the intact, extensive and well-excavated Late Minoan III Necropolis of Armenoi.

The Minoan culture, situated on the Greek island of Crete, has often been called Europe's first civilisation. The people had a structured society with a sophisticated material culture, an intricate administrative system with written records, and far-reaching trade links. The Mycenaean culture, which emerged on mainland Greece ca. 1700 BC, was impacted enormously by the Minoans, but circumstances changed in the Late Minoan III period. Archaeological evidence in Crete reflects a change in burial style, cultural material and language at this time, all of which were linked to Mycenaean influence. However, whether this change was the result of large scale invasion, cultural assimilation or admixture between the two groups is still widely debated (Hallager 2010).

The Late Minoan IIIA–B (ca. 1390–1190 BC) Necropolis of Armenoi is outstanding. As well as being the only intact necropolis from Late Minoan III Crete presently known, the preservation and quality of the finds, and the fully recorded human skeletal remains, are exceptional. Excavation of the 232 tombs excavated to date, containing over 1000 individuals, provided for the largest single sampling of human skeletal remains from the Aegean Bronze Age. As discussed elsewhere in this volume, the diversity of grave goods indicates the possible existence of status differences between buried individuals and tomb groups. The grave goods (principally ceramics and metalwork) are in a Minoan style and yet the burial pattern is new to Minoan Crete, reflecting more the Mycenaean than the Minoan world. Are the people interred here 'invading Mycenaeans' who have adopted the trappings of Minoan civilisation, or Minoans who have adapted to the new Minoan/Mycenaean world, or perhaps, which might be more likely, a mixture of intermarried Minoans and Mycenaeans?

The opportunity to apply scientific techniques to an assemblage that spans one of the key cultural transitions in early European history is unique. Due to the wealth of material, Armenoi can provide us with an in-depth view of Late Minoan society. Our project is the first time that combined DNA analysis, isotopic dietary techniques (see Chapter 4 above for Richards' earlier isotopic analyses) and dating have been applied to a large well-curated burial assemblage, and it is our emphasis on using these different disciplines (genetics, stable isotopic analysis, radiocarbon dating and archaeology) that is the major strength of our research. Our approach will be uniquely informative, with the goal being to answer some of the questions that arose due to the changes that took place in Late Minoan Crete.

Materials and methods

Sample selection

Sample selection was undertaken in February 2017 directly from the storage collection at the Archaeological Museum of Rethymnon, Crete. Petrous bones and molar teeth were preferentially selected due to the increased possibility of DNA

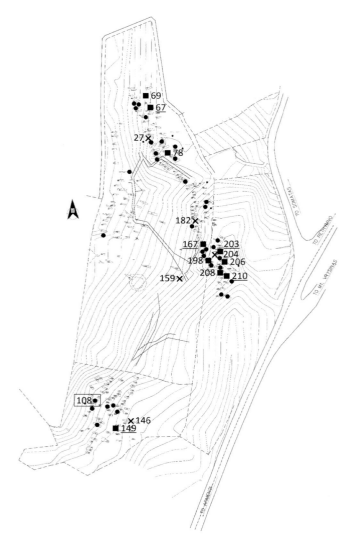

Figure 5.1. Plan of the Necropolis, showing the location of all 48 tombs sampled (as listed in Table 5.1). The 55 individuals analysed in this study came from 16 of these tombs, and tomb numbers are indicated. The remaining 32 tombs are indicated by black dots. Of the 16 tombs analysed, the single tomb that yielded neither stable isotopic nor DNA data (Tomb 108) is enclosed in a box, while those that generated stable isotope results alone are denoted by crosses. The ten tombs that contained the 23 samples analysed by NGS (as listed in Table 5.2) are shown as black squares; these tombs also generated stable isotopic results, and five (underlined) have associated radiocarbon dates (Table 5.3) (the underlying original plan was produced by Steve Litherland and Philip Mann).

and collagen survival (Pinhasi *et al.* 2015; Hansen *et al.* 2017). Elements from broken and fragmented skulls were selected so as not to damage intact skulls, thus allowing the possibility for future research such as anthropological and palaeopathological studies. Although not a precise correlation, the better preserved the bone or enamel matrix

is, the less likely the DNA/collagen inside will have been subjected to post-mortem degradation, especially microbial attack. Therefore, hard, non-flaky petrous bones and teeth without any obvious cracks were selected where possible. Boxes in the museum were arranged by tomb, meaning that, in a number of cases, several individuals were placed in the same box. In boxes containing multiple fragmented skulls the same element was chosen from each individual (for example, the left petrous) to avoid sampling the same individual twice. A variety of tombs with varying wealth (as designated by associated grave goods) was selected from across the site to assess the possibility of status related differences that could be analysed isotopically and/or genetically. In addition, several skeletons were selected from multiple burial tombs to study the possibility of familial relationships.

We wished to sample from Tomb 146, an archaeologically significant tomb containing a stirrup jar used for wine storage, which had the name *wi-na-jo* inscribed in Linear B (Tzedakis and Kolivaki 2018; see Fig. 9.38). As no petrous bones were available from this tomb, we selected both a talus (ankle) bone and an incisor tooth from one of the three individuals interred therein. The only other sample that was not either a petrous or tooth came from Tomb 159, from which we sampled an axis vertebra and a lower left M1 tooth from one of the individuals. Two other individuals were sampled from this tomb, which is the largest at the Necropolis and includes several important archaeological features, such as a column and niches (Tzedakis and Kolivaki 2018; see Chapters 3 and 9).

As shown in Table 5.1, a total of 171 skeletal elements was selected, representing 118 individuals, of which:

- 51 individuals were represented by both petrous and tooth
- 54 individuals were represented by petrous only
- 11 individuals were represented by a tooth only
- 1 individual had a talus and a tooth (from Tomb 146)
- 1 individual had a vertebra and a tooth (from Tomb 159)

These burials come from 48 of the 232 tombs at the Necropolis (Fig. 5.1), including both single occupancy and multi-use tombs with between one and nine people sampled from each tomb. This comprises 34% of the total tombs that have skeletal material. Our selected samples come from a mixture of 'finds poor' tombs (with few/no grave goods) and 'finds rich' tombs (including the largest tomb at the site, Tomb 159, where three of the five burials were suitable for analysis).

Fifty-five of the 118 individuals sampled were analysed in detail as part of this current study. The four individuals taken from Tombs 146 and 159 were chosen, as these are the two most important tombs at the site. In addition, 51 other individuals were selected based on picking the samples with the best morphological preservation. These 55 samples came from 16 tombs, with 1–8 samples taken from each tomb (Table 5.1; Fig. 5.1).

Table 5.1. List of all 171 skeletal fragments selected for analysis, totaling 118 individuals from 48 tombs.

Tomb	Chronology	MNI	Armenoi Project # (envelope #)	Skeleton ID	Element	Endogenous DNA content (%)	$\delta^{13}C$ VPDB	$\delta^{15}N$ AIR	C:N
5	IIIA	5?	2159 (env. 101)	2	right petrous	–	–	–	–
			2160 (env. 101)		upper right M2	–	–	–	–
10*	IIIA2–B2	11	2161 (env. 102)	A	right petrous	–	–	–	–
24**	A2–B	2	2162 (env. 103)	B	lower right M2	–	–	–	–
27	IIIA	4	2021 (env. 5)	1 JUV	left petrous	0.04	poor collagen preservation		
			2022 (env. 6)	1 ADULT	right petrous	0.05	–20.01	9.67	3.3
31	IIIA2end	3	2163 (env. 104)	2	right petrous	–	–	–	–
32	IIIA–B	3	2164 (env. 105)	B	right petrous	–	–	–	–
			2165 (env. 105)		upper right M2	–	–	–	–
67**	IIIA1–B1	11	**2088 (env. 51)**	E	left petrous	26.93	–19.59	10.05	3.2
			2089 (env. 51)		upper right M2	–	–	–	–
			2090 (env. 52)	H	left petrous	0.46	–20.23	9.04	3.5
69	IIIB1	5	**2091 (env. 53)**	Γ	right petrous	4.48	–20.26	8.22	3.5
			2092 (env. 53)		upper left M1	–	–	–	–
78	IIIA1/A2–B1	9	2093 (env. 54)	Z	left petrous	13.48	–19.76	9.01	3.3
			2094 (env. 54)		upper right M2	–	–	–	–
93	IIIA1–B2	7	2166 (env. 106)	Δ	right petrous	–	–	–	–
			2167 (env. 107)	H	right petrous	–	–	–	–
94	IIIA1	9	2168 (env. 108)	ΣΤ	right petrous	–	–	–	–
			2169 (env. 109)	no ID	right petrous	–	–	–	–
100	IIIA2–B1	5	2172 (env. 111)	B	left petrous	–	–	–	–
			2173 (env. 111)		lower left M1	–	–	–	–
			2174 (env. 112)	Δ	right petrous	–	–	–	–
			2170 (env. 110)	Z	left petrous	–	–	–	–
			2171 (env. 110)		upper left M2	–	–	–	–
108	IIIA2–B	11	2095 (env. 55)	B	right petrous	1.73	poor collagen preservation		
			2097 (env. 57)	Δ	right petrous	–	–	–	–
			2098 (env. 57)		lower left M2	–	–	–	–
			2096 (env. 56)	Θ	left petrous	–	–	–	–
114	IIIA2end	4	2099 (env. 58)	B	right petrous	–	–	–	–
			2100 (env. 59)	Γ	right petrous	–	–	–	–
			2101 (env. 60)	Δ	right petrous	–	–	–	–
121	III	4	2175 (env. 113)	no ID	right petrous	–	–	–	–
			2176 (env. 113)		upper right M2	–	–	–	–
132	IIIA2/B1, B1	2	2102 (env. 61)	A	right petrous	–	–	–	–
			2103 (env. 61)		upper left M1	–	–	–	–
133	IIIA2/B1, B1	5	2104 (env. 62)	Δ	left petrous	–	–	–	–
			2105 (env. 62)		lower left M2	–	–	–	–
			2106 (env. 63)	E	right petrous	–	–	–	–
			2107 (env. 63)		lower right M1?	–	–	–	–
146	IIIA–B	3	2014 (env. 1)	A	R talus	0.84	–19.31	9.31	3.3
			2015 (env. 1)		lower right incisor	–	–	–	–

(Continued)

Table 5.1. List of all 171 skeletal fragments selected for analysis, totaling 118 individuals from 48 tombs. (Continued)

Tomb	Chronology	MNI	Armenoi Project # (envelope #)	Skeleton ID	Element	Endogenous DNA content (%)	$\delta^{13}C$ VPDB	$\delta^{15}N$ AIR	C:N
149	IIIA2–B1	6	2030 (env. 11)	A	right petrous	0.29	poor collagen preservation		
			2025 (env. 8)	Δ	left petrous	9.93	−19.72	9.42	3.3
			2026 (env. 8)		upper right M3?	–	–	–	–
			2028 (env. 10)	E	left petrous	3.70	poor collagen preservation		
			2029 (env. 10)		lower left M2?	–	–	–	–
			2023 (env. 7)	H	right petrous	0.30	−20.46	9.12	3.4
			2024 (env. 7)		upper left M3	–	–	–	–
			2027 (env. 9)	ΣΤ	right petrous	7.01	−19.52	9.53	3.2
159**	IIIA1/2–B2	5	2016 (env. 2)	A	left petrous	0.23	−20.04	7.70	3.3
			2017 (env. 2)		upper right M2	–	–	–	–
			2020 (env. 4)	Γ	lower left M3	0.28	−19.33	10.62	3.2
			2018 (env. 3)	Δ	axis vertebra	–	–	–	–
			2019 (env. 3)		lower left M1	0.30	no collagen		
160	IIIA2	5	2108 (env. 64)	Δ	left petrous	–	–	–	–
161	IIIA2–B1	7	2109 (env. 65)	ΣΤ	right petrous	–	–	–	–
162	IIIA2– B1late	8	2110 (env. 66)	Z	right petrous	–	–	–	–
			2111 (env. 66)		lower left M2	–	–	–	–
167**	IIIA1/2–B1middle	7	2032 (env. 13)	B	upper left M3	0.27	−19.02	9.51	3.2
			2035 (env. 15)	Δ	upper left M2	0.41	−19.76	7.58	3.3
			2036 (env. 16)	E	left petrous	4.12	−19.51	8.86	3.2
			2037 (env. 16)		lower left M2	–	–	–	–
			2033 (env. 14)	Z	right petrous	2.91	poor collagen preservation		
			2034 (env. 14)		lower left M2	–	–	–	–
			2031 (env. 12)	ΣΤ	right petrous	1.33	−19.57	9.39	3.2
173	IIIA2–B1middle	7	2112 (env. 67)	A	right petrous	–	–	–	–
			2114 (env. 69)	E	left petrous	–	–	–	–
			2113 (env. 68)	Σ	right petrous	–	–	–	–
174	IIIA1– A2/B1	2	2115 (env. 70)	B	right petrous	–	–	–	–
			2116 (env. 70)		upper right M1	–	–	–	–
175	IIIA1/2	1	2117 (env. 71)	Γ	right petrous	–	–	–	–
			2118 (env. 71)		upper right M2	–	–	–	–
177	IIIA2–B1	8	2124 (env. 75)	A	right petrous	–	–	–	–
			2125 (env. 75)		upper right M1	–	–	–	–
			2126 (env. 76)	B	left petrous	–	–	–	–
			2127 (env. 77)	Γ	left petrous	–	–	–	–
			2128 (env. 77)		upper left M2	–	–	–	–
			2120 (env. 73)	E	right petrous	–	–	–	–
			2121 (env. 73)		upper left M2	–	–	–	–
			2122 (env. 74)	H	right petrous	–	–	–	–
			2123 (env. 74)		upper right M2	–	–	–	–
			2119 (env. 72)	2	right petrous	–	–	–	–
178	IIIA–A2/B	2	2129 (env. 78)	A	left petrous	–	–	–	–

(Continued)

Table 5.1. (Continued)

Tomb	Chronology	MNI	Armenoi Project # (envelope #)	Skeleton ID	Element	Endogenous DNA content (%)	$\delta^{13}C$ VPDB	$\delta^{15}N$ AIR	C:N
179*	IIIA1–B1	9	2130 (env. 79)	A	right petrous	–	–	–	–
			2131 (env. 80)	Z	right petrous	–	–	–	–
			2132 (env. 80)		upper right M2	–	–	–	–
184*	IIIA1/2–A2/B1	12	2136 (env. 83)	2	right petrous	–	–	–	–
			2137 (env. 84)	3	right petrous	–	–	–	–
			2133 (env. 81)	7	right petrous	–	–	–	–
			2134 (env. 82)	8	right petrous	–	–	–	–
			2135 (env. 82)		lower left M2	–	–	–	–
187	IIIA1–B1	3	2177 (env. 114)	1	right petrous	–	–	–	–
			2178 (env. 114)		upper left M2	–	–	–	–
			2179 (env. 115)	2	upper right M2	–	–	–	–
188	IIIA1–B1	8	2138 (env. 85)	A	right petrous	–	–	–	–
			2139 (env. 86)	1	left petrous	–	–	–	–
			2140 (env. 86)		upper left M1	–	–	–	–
			2141 (env. 87)	3	right petrous	–	–	–	–
189	IIIA–B1	11	2038 (env. 17)	2	left petrous	0.07	–19.80	9.37	3.3
			2039 (env. 17)		upper right M3	–	–	–	–
			2040 (env. 18)	11	left petrous	0.05	poor collagen preservation		
190	IIIA2/B1–B	10	2142 (env. 88)	B	left petrous	–	–	–	–
			2143 (env. 89)	Δ	left petrous	–	–	–	–
			2144 (env. 90)	I	left petrous	–	–	–	–
198*	IIIA2–B2	8	2046 (env. 22)	B	right petrous	12.96	–19.90	9.47	3.3
			2041 (env. 19)	Γ	left petrous	25.43	–20.54	9.03	3.5
			2042 (env. 19)		upper right M2	–	–	–	–
			2043 (env. 20)	Δ	upper left M2	13.02	–19.48	8.73	3.2
			2044 (env. 21)	ΣΤ	left petrous	0.11	–19.49	10.71	3.3
			2045 (env. 21)		upper right M1	–	–	–	–
			2047 (env. 23)	no ID	lower right M2	0.13	–19.81	8.98	3.3
199	IIIA1late–A2	9	2147 (env. 92)	B	left petrous	–	–	–	–
			2148 (env. 92)		upper left M1	–	–	–	–
			2145 (env. 91)	Γ	right petrous	–	–	–	–
			2146 (env. 91)		upper left M2	–	–	–	–
201	IIIA1–A2late	12	2151 (env. 94)	Δ	right petrous	–	–	–	–
			2149 (env. 93)	ΣΤ	right petrous	–	–	–	–
			2150 (env. 93)		upper left M2	–	–	–	–
203*	IIIA1–B	11	2053 (env. 28)	A	left petrous	2.68	–19.65	9.84	3.3
			2054 (env. 28)		upper left M2	–	–	–	–
			2049 (env. 25)	E	left petrous	3.99	–19.69	9.69	3.2
			2050 (env. 25)		upper left M2	–	–	–	–
			2056 (env. 30)	Z	left petrous	9.12	–19.65	9.94	3.3
			2055 (env. 29)	H	right petrous	5.55	–20.49	8.95	3.3

(Continued)

Table 5.1. List of all 171 skeletal fragments selected for analysis, totaling 118 individuals from 48 tombs. (Continued)

Tomb	Chronology	MNI	Armenoi Project # (envelope #)	Skeleton ID	Element	Endogenous DNA content (%)	$\delta^{13}C$ VPDB	$\delta^{15}N$ AIR	C:N
			2048 (env. 24)	Θ	right petrous	14.79	−19.86	10.13	3.4
			2052 (env. 27)	I	right petrous	7.11	−19.83	9.30	3.3
			2057 (env. 31)	K	right petrous	16.47	−19.58	9.49	3.4
			2051 (env. 26)	ΣΤ	left petrous	6.93	−20.05	8.80	3.3
204	IIIA2, A2/B1	8	2058 (env. 32)	A	right petrous	0.28	−19.82	8.57	3.5
			2059 (env. 33)	E	right petrous	0.14	−19.89	10.34	3.4
			2060 (env. 33)		upper left M2	–	–	–	–
			2061 (env. 34)	Z	right petrous	17.18	−20.01	8.82	3.4
206**	IIIA1–B2	16	2066 (env. 38)	A	right petrous	0.01	poor collagen preservation		
			2067 (env. 38)		lower left M2	–	–	–	–
			2072 (env. 41)	B	right petrous	0.01	poor collagen preservation		
			2073 (env. 41)		upper left M2?	–	–	–	–
			2064 (env. 36)	Δ	upper right M2	0.04	−19.24	8.30	3.2
			2062 (env. 35)	Z	left petrous	0.07	poor collagen preservation		
			2063 (env. 35)		upper left M2	–	–	–	–
			2068 (env. 39)	I	left petrous	0.28	−19.68	9.74	3.3
			2069 (env. 39)		upper left M2	–	–	–	–
			2076 (env. 43)	M	left petrous	0.02	poor collagen preservation		
			2074 (env. 42)	O	left petrous	0.07	−19.89	8.79	3.3
			2075 (env. 42)		upper left M3?	–	–	–	–
			2065 (env. 37)	ΣΤ	left petrous	0.04	poor collagen preservation		
			2070 (env. 40)	no ID	left petrous	0.02	poor collagen preservation		
			2071 (env. 40)		upper right M2	–	–	–	–
208	IIIA1, A2/B1	6	2077 (env. 44)	B	right petrous	0.02	poor collagen preservation		
			2078 (env. 44)		upper right M2	–	–	–	–
			2079 (env. 45)	E	right petrous	5.17	−20.25	8.81	3.3
209	IIIB	4	2152 (env. 95)	Γ	right petrous	–	–	–	–
			2153 (env. 95)		upper right M1	–	–	–	–
210	IIIA2–B	11	2084 (env. 48)	E	right petrous	2.57	−20.69	7.81	3.2
			2087 (env. 50)	1	left petrous	0.70	−20.16	8.75	3.2
			2085 (env. 49)	2	left petrous	0.42	−20.05	8.20	3.2
			2086 (env. 49)		upper left M2?	–	–	–	–
			2080 (env. 46)	3	left petrous	0.04	−19.64	8.04	3.2
			2081 (env. 46)		lower left M2	–	–	–	–
			2082 (env. 47)	4	right petrous	11.45	−20.11	9.15	3.3
			2083 (env. 47)		lower right M1	–	–	–	–
211*	IIIA2, A2/B1	10	2154 (env. 96)	A/B	left petrous	–	–	–	–
218	IIIA2/B	5	2155 (env. 97)	A	lower left M2	–	–	–	–
			2158 (env. 100)	Δ	upper left M1	–	–	–	–
			2156 (env. 98)	Δ-1	left petrous	–	–	–	–
			2157 (env. 99)	Δ-2	left petrous	–	–	–	–

(Continued)

Table 5.1. (Continued)

Tomb	Chronology	MNI	Armenoi Project # (envelope #)	Skeleton ID	Element	Endogenous DNA content (%)	$\delta^{13}C$ VPDB	$\delta^{15}N$ AIR	C:N
220	unknown	2	2180 (env. 116)	A	left petrous	–	–	–	–
			2181 (env. 116)		lower right M2	–	–	–	–
			2182 (env. 117)	B	right petrous	–	–	–	–
			2183 (env. 117)		lower right M1	–	–	–	–
229	IIIA2	2	2184 (env. 118)	A	upper right M2	–	–	–	–

Tombs with an asterisk (*) are 'finds rich', with two asterisks (**) denoting the wealthiest tombs. MNI denotes the mean number of individuals from each tomb. We aimed to take as many people from each tomb as possible, and the individual samples listed are those where the skulls were not intact, where petrous or teeth (including one incisor, plus molars M1, M2 and M3) remained (apart from in the case of Tomb 146 as mentioned in the text), and where the preservation was deemed suitable for bioarchaeological analyses. The 55 individuals, from 16 tombs, who were screened for DNA and underwent stable isotopic analysis, are noted, with the corresponding endogenous DNA content listed in percentage, and the bone/tooth root collagen dietary stable isotope data. The Armenoi Project numbers in bold denote the eight samples that were sent for radiocarbon dating (Table 5.3).

Dietary stable isotopic analysis of 55 human and 33 animal bones

As indicated by Richards and Hedges (2008 and see Chapters 4 and 8), it is essential to co-analyse faunal remains that represent possible food sources for the people buried at the Necropolis for the purposes of interpretation of stable isotopic data. Richards and Hedges (2008) created their faunal baseline using animals from the MM I–LM IIIC (ca. 2160–1190 BC) site of Chamalevri, 8 km to the north-east of the Necropolis. As differences in local foddering and manuring can affect dietary isotopic measurements, leading to large variation within a small region (Nitsch *et al.* 2017), we created a more local and temporal animal isotopic baseline to aid data interpretation. Remains from 13 animals, which had been placed inside the tombs themselves, were taken from the Necropolis, with one bone from Tomb 146 and 12 from Tomb 159. Results of the species identification of these bones, using both standard zooarchaeological assessment and ZooMS analyses, are given in the Appendix (Table 5.5). In addition, the excavators provided material from 20 faunal remains (ten cattle, eight sheep/goat and two pig) from the Bee Garden excavation of the town site situated on the top of the hill above the Necropolis, where the modern village of Kastellos is now located (see Chapter 9). The lowest levels of the town have been securely dated to LM IIIA1, the same date as the earliest finds from the chamber tombs at the Necropolis.

Bone collagen was extracted from the 55 humans, as well as from the 33 animals, following the modified Longin (1971) method. Carbon and nitrogen stable isotopic (SI) analyses were undertaken at the University of Oxford on a Sercon 2022/Europa GSL-EA continuous flow isotope ratio mass spectrometer (CF-IRMS) system, together with alanine standards to correct for machine drift, and replicates of two internal standards (cow and seal bone collagen) to calibrate values and correct for scale-compression effects (Coplen *et al.* 2006). The alanine, cow and seal bone collagen standards are regularly checked against USGS40 and USGS41 glutamic acid standards and are thus traceable back to VPDB and AIR scales for carbon and nitrogen stable isotopic composition respectively. The consensus values for these three standards are −27.11‰ and −1.56‰ for alanine, −24.3‰ and 8.0‰ for cow and −12.0‰ and 16.6‰ for seal, for $\delta^{13}C$ (VPDB) and $\delta^{15}N$ (AIR) respectively. The actual mean values obtained during the numerous instrument runs for the analyses reported here were −27.09‰ and −1.50‰ for alanine, −24.3‰ and 8.0‰ for cow and −12.0‰ and 16.6‰ for seal. In each case, standard deviations were less than 0.1‰ for $\delta^{13}C$ and less than 0.2‰ for $\delta^{15}N$.

Ancient DNA analyses of 55 human samples

Extractions of the 55 human samples were undertaken in the dedicated Ancient DNA Facility at the University of Huddersfield, which has complete physical separation from where modern DNA products are handled and analysed. Within this carefully restricted suite of clean-room laboratories, surfaces, dedicated equipment and reagents are decontaminated with bleach, ethanol and UV light exposure, while a positive air pressure system within the building reduces the introduction of external contaminants. External layers of each bone or tooth were UV-irradiated, air-abraded, sectioned and pulverised. The resulting powder was incubated with proteinase buffer, concentrated using a 30kDa filter, and silica-column purified, following the extraction protocol by Yang *et al.* (1998) with modifications by MacHugh *et al.* (2000). Blank controls were included throughout the sampling procedure, extraction and library preparation steps to allow for estimation of modern DNA contamination.

Next-generation sequencing (NGS) libraries were constructed for shotgun screening using the method by Meyer and Kircher (2010), with modifications outlined in Gamba *et al.* (2014) and Martiniano *et al.* (2014). All 55 samples were screened by Valeria Mattiangeli (Trinity College Dublin) using a MiSeq to assess DNA quality, and the endogenous DNA content was calculated in-house at Huddersfield. The samples with an endogenous DNA content of greater than 2.5% (Table 5.1) and displaying damage patterns consistent with ancient DNA (not shown), were selected for further in-depth NGS. This equated to 23 individuals from ten different tombs, with 1–8 samples taken from each tomb (Table 5.1; Fig. 5.1). Between one and three libraries were constructed for each sample, as described above. However, unlike the screening libraries, a USER (Uracil-Specific Excision Reagent) enzyme step

was included to repair some of the deamination damage inherent in the ancient DNA, and thus generate the most useable sequence data from the endogenous DNA as possible. The resulting libraries were sent to Macrogen, South Korea, for shot-gun NGS on nine lanes of an Illumina HiSeq4000 machine.

Although experienced osteoarchaeologists working with complete pelvises and/or skulls have a 90–95% accuracy in determining sex, it is not always possible to determine the sex of incomplete or juvenile skeletons (Mays 2010). For this reason, the methods outlined by Skoglund *et al.* (2013) were used to determine the genetic sex of the 23 individuals. Kinship analysis was also undertaken, using the program READ (Relationship Estimation from Ancient DNA; Kuhn *et al.* 2018), which designates pairs of individuals as being first-degree, second-degree, or unrelated.

Table 5.2. List of the 23 samples that had endogenous DNA content greater than 2.5% (Table 5.1) and were sent for shot-gun NGS analysis.

Tomb.	Chronology	MNI	Armenoi Project # (envelope #)	Skeleton ID	Element sampled	Endogenous DNA content (%)	Genetic sex	mtDNA	Y
67**	IIIA1–B1	11	2088 (env. 51)	E	L petrous	26.93	XY	H1bz	I1 or J2
69	IIIB1	5	2091 (env. 53)	Γ	R petrous	4.48	XX	H	–
78	IIIA1/A2–B1	9	2093 (env. 54)	Z	L petrous	13.48	XY	HV1	G2a
149	IIIA2–B1	6	2025 (env. 8)	Δ	L petrous	9.93	XX	H59	–
			2028 (env. 10)	E	L petrous	3.70	XY	T1a	R1b
			2027 (env. 9)	ΣT	R petrous	7.01	XX	T2b	–
167**	IIIA1/2–B1 middle	7	2036 (env. 16)	E	L petrous	4.12	XX	U7b	–
			2033 (env. 14)	Z	R petrous	2.91	XX	U7	–
198*	IIIA2–B2	8	2046 (env. 22)	B	R petrous	12.96	XX	H1e	–
			2041 (env. 19)	Γ	L petrous	25.43	XX	H1e	–
			2043 (env. 20)	Δ	ULM2	13.02	XX	H4b	–
203*	IIIA1–B	11	2053 (env. 28)	A	L petrous	2.68	XX	H5a3	–
			2049 (env. 25)	E	L petrous	3.99	XY	U5b	G2a
			2056 (env. 30)	Z	L petrous	9.12	XX	U5b	–
			2055 (env. 29)	H	R petrous	5.55	XX	U5b	–
			2048 (env. 24)	Θ	R petrous	14.79	XX	U5b	–
			2052 (env. 27)	I	R petrous	7.11	XY	U5a	G2a
			2057 (env. 31)	K	R petrous	16.47	XY	U5a	G2a
			2051 (env. 26)	ΣT	L petrous	6.93	XX	K1	–
204	IIIA2, A2/B1	8	2061 (env. 34)	Z	R petrous	17.18	XX	N1a	–
208	IIIA1, A2/B1	6	2079 (env. 45)	E	R petrous	5.17	XY	H59	J
210	IIIA2–B	11	2084 (env. 48)	E	R petrous	2.57	XX	W	–
			2082 (env. 47)	4	R petrous	11.45	XX	W	–

These individuals come from ten tombs, with between one to eight people sampled from each. The majority of the DNA was extracted from the petrous bone. The genetic sex (XX = female and XY = male), and mitochondrial and Y-chromosome (where applicable) haplogroups are noted for each individual. Tombs with an asterisk (*) are 'finds rich', with two asterisks (**) denoting the wealthiest tombs.

In order to study ancient population structure, a Principal Components Analysis (PCA) plot was generated using Smartpca v.16000 from EIGENSOFT (Patterson *et al.* 2006). Modern genetic variation was sampled from the Human Origins Project (Patterson *et al.* 2012; Lazaridis *et al.* 2014; 2016) and published data from modern Greece was also included. Ancient Greek samples (Lazaridis *et al.* 2017; Stamatoyannopoulos *et al.* 2017; Mathieson *et al.* 2018; Drineas *et al.* 2019) and our 23 Necropolis samples were then projected onto this background. ADMIXTURE v.1.3.0 (Alexander *et al.* 2009), which calculates the different ancestry components present, was employed using a dataset of relevant modern and ancient populations (Lazaridis *et al.* 2017).

Radiocarbon dating

The Necropolis of Armenoi dates to the Late Minoan period, LM IIIA–B, which equates to 1390–1190 BC, a period of 200 years. The tombs were dated contextually on the basis of ceramic typology. However, as many tombs are multi-use and may have been used over several generations, we wanted to test if radiocarbon dates might allow a more detailed timeline of the site, aiding interpretation of the DNA results and providing a better understanding of how long tombs were in use. Therefore, eight samples were sent to ¹⁴Chrono, the radiocarbon dating laboratory at Queen's

University Belfast. All samples were from individuals that had been sent for in-depth NGS, and the reasons behind each choice is given in Table 5.3. Of these, six returned radiocarbon dates, which were calibrated using OxCal version 4.4.3 (Bronk-Ramsey 2009) and the most recent calibration curve, IntCal20 (Reimer *et al.* 2020).

As the outlier individual from Tomb 149 (Skeleton E) failed with conventional dating methods, a sample was also sent to ORAU (the University of Oxford Radiocarbon Accelerator Unit) for further analysis. In order to determine the yield of raw collagen, a 750 mg sub-sample was extracted using both the ultra-filtration (AF) and gelatin (AG) methods (Brock *et al.* 2010).

Dietary stable isotopes
Basic principles

Dietary stable isotope analysis assesses the ratios of $^{13}C/^{12}C$ ($\delta^{13}C$) and $^{15}N/^{14}N$ ($\delta^{15}N$). As carbon and nitrogen from food are incorporated into living tissue, their isotopic ratios can be used for dietary reconstruction. Body tissues that are continually remodelled during life, such as bone, have isotopic signatures that represent the average diet during the last few years of the person's life. However, it must be remembered that diet is not only determined by the food available but also by culture (Twiss 2012).

Table 5.3. Radiocarbon results from the eight individuals sent for dating, with associated reasons for selection.

Sample ID	Armenoi Project no. (envelope no)	Reason for dating	Lab. ref.	Date (BP)	Calibrated date (cal BC)
Tomb 67** E	2088 (env. 51)	rich tomb at north of site	UBA-44547	3131±33	**1497**–1471 (6.4%)
					1464–1369 (63.3%)
					1356–**1297** (25.8%)
Tomb 69 Γ	2091 (env. 53)	poor tomb at north of site	UBA-44548	low collagen yield	n/a
Tomb 149 Δ	2025 (env. 8)	same tomb as outlier	UBA-43710	3025±32	**1396**–1332 (25.8%)
					1326–1196 (66.6%)
					1174–1163 (1.4%)
					1143–**1131** (1.6%)
Tomb 149 E	2028 (env. 10)	West European outlier	UBA-43712 P50133	no collagen	n/a
Tomb 149 ΣT	2027 (env. 9)	same tomb as outlier	UBA-43711	3100±28	1432–1282 (95.4%)
Tomb 167** E	2036 (env. 16)	central tomb	UBA-44544	3095±30	1429–1274 (95.4%)
Tomb 203* I	2052 (env. 27)	'oldest' member of family group	UBA-44545	3151±30	**1500**–1384 (87.7%)
					1340–**1315** (7.7%)
Tomb 210 4	2082 (env. 47)	most southerly of the central tombs with DNA	UBA-44546	3080±30	1421–1263 (95.4%)

The calibration is 2σ undertaken using OxCal (Bronk Ramsey 2009) and the most recent calibration curve, IntCal20 (Reimer *et al.* 2020). As before, tombs with an asterisk (*) are 'finds rich', with two asterisks (**) denoting the wealthiest tombs. All the date ranges broadly fit within the typology contextual dates for the Necropolis of 1390–1190 BC

Differing $\delta^{13}C$ values are a result of the two photosynthetic pathways, which in turn lead to differences in isotopic fractionation, with more positive $\delta^{13}C$ being found in C4 plants (e.g., millet), which live in more arid environments and have a pathway that conserves more moisture, than in C3 plants, such as cereals. Carbon can also be used to differentiate terrestrial from marine based diets. The carbon in terrestrial food chains mainly derives from atmospheric carbon dioxide whereas, in marine food chains, the main source of carbon is dissolved bicarbonate, which is enriched for $\delta^{13}C$ when compared to atmospheric carbon. Therefore, diets that are more reliant on marine sources exhibit more positive carbon values than those reliant on terrestrial C3 plants (Schulting 1998; Lamb *et al.* 2014).

The nitrogen isotope ratio, $\delta^{15}N$, reflects the trophic level of the consumer (*i.e.*, the hierarchical position of the organism in the local food chain). There is an increase of 3–5‰ between plants and the herbivores that consume them, and a similar increase between herbivores and their predators, which can lead to carnivores in long food chains having very high $\delta^{15}N$ values. For this reason, higher $\delta^{15}N$ values are often associated with the longest food chains, such as those typically found in aquatic (particularly marine) ecosystems. However, the $\delta^{15}N$ value can also be affected by aridity, sea spray and manuring (Bogaard *et al.* 2007; Dotsika *et al.* 2019).

Nitrogen and carbon values combined can more accurately differentiate diet. As both are often preserved within the collagen protein fraction of archaeological bone, dietary stable isotopic studies of collagen carbon and nitrogen stable isotopic composition have been used widely to assess diet within skeletal assemblages (Schoeninger and Moore 1992). However, due to differences in the uptake of these isotopes at different locations by the species consumed, it is important to analyse contemporary animal samples in concert with human collagen to create baselines for carbon and nitrogen when assessing human diets (Papathanasiou and Richards 2015; Richards 2015). This can be problematic when dealing with human cemetery sites where faunal material may be scarce or absent.

Dietary stable isotopes from Greece

Apart from some individuals at Grave Circles A and B at Mycenae (16th and 17th centuries BC, respectively), who provided evidence for some marine consumption, people from mainland Greece, including from the later Mycenaean chamber tombs (1600–1200 BC), appear to have eaten a mainly terrestrial diet (Richards and Hedges 2008; Triantaphyllou *et al.* 2008; Papathanasiou 2015). Richards and Hedges (2008) studied 39 individuals from the of Armenoi Necropolis and found a primarily C3 plant based protein source with little/no marine input (see also Chapter 4). They analysed faunal remains from the MM I–LM IIIC (ca. 2160–1190 BC) site of Chamalevri, 8 km to the northeast of Armenoi as a baseline comparison and concluded that the Necropolis population had a high meat consumption with little variation in diet seen across the site, although males had slightly increased nitrogen values when compared to females. In contrast, two sites near Knossos, Ailias chamber tomb 55 and the Lower Gypsades tholos tomb and ossuary, both of which date to MM III–LM I (ca. 1700–1600 BC), showed evidence of both terrestrial and marine diets (Nafplioti 2016).

Clear dietary differences have also been associated with both status and sex at several sites on the Greek mainland, with elites having enriched nitrogen isotopes (Papathanasiou and Richards 2015). In 2019, to examine dietary changes over time, Dotsika *et al.* reviewed stable isotopic data of mainland Greek samples from 22 sites dating from the Early Neolithic to the Late Bronze period. Their results supported previous studies, with diets being predominantly terrestrial C3 based. However, a small amount of marine consumption was identified at the Neolithic sites and two of the Early Bronze Age sites, one of which, Perachora near Corinth, was described as having a mix of terrestrial and marine. The Late Bronze site of Voudeni, near Patras in the north-west Peloponnese, also had a marine input, but once again not a significant amount.

Results and discussion of our dietary stable isotope analyses

Of the 33 animal samples from the Necropolis and nearby town site, all produced collagen for analysis (Table 5.4). However, sheep ARM01 had a C:N ratio value of 4.6, which, as higher than the range of values expected from diagenetically unaltered mammalian collagen (of between 2.9 and 3.6; DeNiro 1985; Ambrose 1990; van Klinken 1999), was excluded from further analysis. Dietary isotopic results from the animals located at the town site had higher $\delta^{15}N$ that those from Chamalevri (Richards and Hedges 2008). However, the sheep/goat samples from Tomb 159 of the Necropolis are broadly compatible with the values reported by Richards and Hedges (2008). These differences between sites are likely to reflect local variation in grazing practices and/or environmental differences.

Of the humans, 41 of the 55 produced enough well-preserved collagen to be analysed (Table 5.1). Poor preservation of collagen did not appear to have a linear relationship to endogenous DNA survival. However, of the 14 samples where collagen could either not be extracted, or where the C:N ratio was outside of the expected range, 12 had an endogenous DNA content of <2.5%, which was our cut-off for considering them for in-depth sequencing.

Table 5.4. Faunal bone collagen dietary stable isotope data from Tomb 159 and the nearby town site.

Species	Code	Element sampled	δ¹³C VPDB	δ¹⁵N AIR	C:N
		Tomb 159			
sheep	ARM02	metatarsal	−20.36	5.38	3.3
sheep	ARM05	humerus	−21.20	4.85	3.4
sheep	ARM06	femur	−21.31	3.85	3.4
sheep	ARM07	femur	−21.07	4.81	3.4
sheep	ARM08	humerus	−21.35	4.21	3.4
sheep	ARM12	tibia	−20.94	5.41	3.3
sheep	ARM13	mandible	−20.82	5.04	3.4
goat	ARM03	metatarsal	−20.26	3.36	3.3
hare	ARM04	femur	−22.64	1.86	3.4
hare	ARM09	humerus	−20.90	2.98	3.4
hare	ARM10	humerus	−22.56	2.99	3.4
hare	ARM11	humerus	−20.83	3.80	3.3
		Armenoi town site			
sheep/ goat	TOMH1	phalanx	−23.40	5.81	3.2
sheep/ goat	TOMH1	metapodial	−22.67	6.05	3.5
sheep/ goat	TOMH2	calcaneus	−22.63	5.89	3.3
sheep/ goat	TOMH2	metapodial	−21.54	6.01	3.3
sheep/ goat	TOMH2	ulna	−21.45	4.89	3.6
sheep/ goat	TOMH2	calcaneous	−22.78	6.84	3.3
sheep/ goat	TOMH3	radius	−22.99	6.22	3.5
sheep/ goat	TOMH3	phalanx	−23.18	5.81	3.1
cattle	TOMH1	mandible	−24.54	6.73	3.3
cattle	TOMH1	mandible	−23.87	7.12	3.4
cattle	TOMH1	metacarpal	−24.51	6.94	3.2
cattle	TOMH1	calcaneum	−23.26	7.66	3.4
cattle	TOMH1	metatarsal	−22.49	7.81	3.2
cattle	TOMH1	phalanx	−24.81	7.32	3.3
cattle	TOMH1	metatarsal	−22.97	7.58	3.4
cattle	TOMH1	mandible	−23.77	6.99	3.2
cattle	ARMEN01	scapula	−23.40	6.85	3.2
cattle	ARMEN01	phalanx	−23.49	8.10	3.5
pig	TOMH1	metapodial	−19.27	9.18	3.4
pig	TOMH1	phalanx	−19.41	8.76	3.4

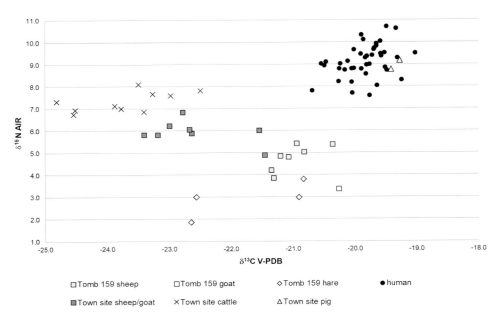

Figure 5.2. Dietary stable isotope plot of human and faunal samples from the LM III Necropolis of Armenoi and faunal samples from the nearby town site. Symbols are shown in the accompanying key.

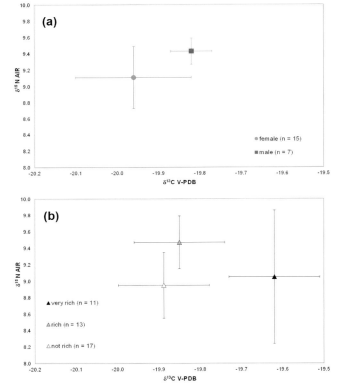

Figure 5.3. Dietary stable isotope plots comparing (a) the male and female samples (n = 22), and (b) individuals from 'very rich', 'rich' and 'not rich' tombs (n = 41).

Overall, the Necropolis population produced δ¹³C values between –19.02‰ and –20.69‰ and δ¹⁵N values between 7.58‰ and 10.71‰. These are far below those expected for a significant input of marine or C4 plant-based

dietary protein, and approximately 1‰ and 3–4‰ higher, respectively, than the animal baselines, which is in line with consumption enrichment as mentioned in Richards and Hedges (2008). The human δ¹³C values are significantly offset (by approximately 3‰) from the town site herbivore values, particularly so for cattle. This is more than can be explained by the trophic level effect for δ¹³C and suggests that the town site fauna were not a significant part of the diet of the human population who were buried at the Necropolis. When compared to the sheep/goat from Tomb 159, it suggests that the herbivores from the town site were grazed in a different, and potentially more wooded or enclosed, environment than the fauna represented at the Necropolis itself.

Overall both the δ¹³C and δ¹⁵N values suggest that the Necropolis population were mainly consuming a relatively homogeneous, terrestrial diet, with C3 plants at the base of the food chain (Fig. 5.2). Although, overall, males were eating a diet higher in δ¹⁵N, this amounts to less than 0.4‰ difference and is not significant (Fig. 5.3a). When tomb wealth is considered, although there is a difference seen between the diets from very rich, rich and other tombs, there is a large degree of overlap between the data from the three wealth categories, with no supporting statistical significance to the observations (Fig. 5.3b). Three juveniles (Tomb 198 ΣT, Tomb 198 no ID, and Tomb 210 Skeleton 4) produced collagen suitable for stable isotopic analysis, and these all plotted alongside the adults, with Tomb 198 ΣT having the highest δ¹⁵N value of the Necropolis, which is likely to represent a pre-weaning/breast-feeding dietary signal in this individual.

The human dietary stable isotopic results presented here (Fig. 5.2) confirm the homogeneous, terrestrial based diet at the site, as determined by Richards and Hedges (2008; see also Chapter 4). This suggests that, despite being close to the sea, hunting and farming were clearly more culturally important. Although the absolute values from the Chamalevri fauna are not reported in Richards and Hedges (2008), the approximate mean values can be estimated from their figure 1, as 4‰ for $\delta^{15}N$ and –20 ‰ for $\delta^{13}C$. These values are significantly different to the faunal values that we obtained from the town site nearby the Necropolis. The increased $\delta^{15}N$ values seen at the town site demonstrate that Crete had a range of local isotopic baselines, as would be expected in such an ecologically diverse setting. If significant amounts of meat from these animals were being consumed, their elevated $\delta^{15}N$ values would have implications for any reconstruction of the amount of meat included in the diet of the Necropolis population. However, our $\delta^{13}C$ results suggest that the people buried at the Necropolis were not sourcing their food from the town site (see below and Chapter 8).

The sheep and goat bones from Tomb 159, with a mean $\delta^{15}N$ value of 4.6‰, are approximately a full trophic level below the mean of the human data, whereas the town site fauna, with the sheep/goat samples having a mean $\delta^{15}N$ value of 5.9‰ and the cattle 7.3‰, are considerably less than a full trophic level below the human values. If these higher $\delta^{15}N$ meat sources from the town site were a factor of the Necropolis diet, considerably less would need to be consumed to produce the observed values in the human collagen. This would lead to the conclusion that the Necropolis population did not have a diet as rich in animal protein as previously assumed (Richards and Hedges 2008). However, the relatively low $\delta^{13}C$ values from the town site fauna measured in this study suggest that these samples were not the source of the animal protein consumed by the cemetery population, as they are considerably more depleted in $\delta^{13}C$ than would be accounted for by a single trophic level shift. Based on the carbon isotopic results, it seems likely that the faunal values measured here from Tomb 159, or from the fauna measured from Chamalevri by Richards and Hedges (2008), are a more appropriate measure of the faunal isotopic base line for the human population of the Armenoi cemetery. If so, this supports the conclusion of Richards and Hedges that the diet of the people represented in the cemetery was relatively rich in terrestrial animal-based protein; *i.e.* meat and/or dairy products (see Chapter 4).

The lack of marine consumption has been seen in several mainland Greek sites but contrasts with Mycenaean elite graves (Richards and Hedges 2008; Papathanasiou and Richards 2015). Mycenaean sites tend to exhibit status based differences, which are not present at the LM III Necropolis of Armenoi (Fig, 5.3b), and we found no significant dietary differences related to sex (Fig. 5.3a); this is in agreement with the findings previously reported by Richard and Hedges (2008). The homogeneous, terrestrial based diet seen at the Necropolis contrasts with that found at the MM III–LM (ca. 1700–1600 BC) tombs at Knossos (Nafplioti 2016), which exhibited a range of diets featuring both terrestrial and marine input; however, the time scale is such (a difference of five or more generations) that no conclusions can be drawn.

Ancient DNA

Basic principles

Deoxyribonucleic acid (DNA) provides the genetic code for life, specifying the biological characteristics of an individual, including biological sex, indicators of phenotype and ancestry information (Jobling *et al.* 2013; Herrera and Garcia-Bertrand 2018). DNA can be inherited by two methods: uniparentally (either from the mother or the father) and biparentally (from both parents). The uniparental markers include the mitochondrial DNA (mtDNA), which is inherited maternally, and the Y-chromosome, which is inherited paternally. For the mtDNA, since it does not recombine, any mutations are passed mother to child, which has allowed the construction of phylogenetic trees of similar sequences known as haplogroups. By convention, haplogroups are given letters, and subhaplogroups are given numbers, then followed by a letter, etc. For example, J1a is a subhaplogroup of J1, which is a subhaplogroup of J. The Y-chromosome has a similar naming onvention, and both uniparental markers can be useful in identifying familial links. Conversely, the autosomal chromosomes (*i.e.* any chromosome that is not a sex chromosome) are inherited from both parents and, therefore, can be studied to inform on ancestry.

When an organism dies, its DNA undergoes post-mortem degradation due to taphonomic factors such as temperature and pH. Other processes, of hydrolysis, oxidation and microbial attack, can cause deamination (leading to miscalling of the sequence), fragmentation and loss of endogenous DNA (*i.e.*, the DNA that belongs to the individual), as well as an increased risk of contamination with modern DNA (Heintzman *et al.* 2015; Marciniak and Perry 2017). As archaeological bones are a finite resource and ancient DNA analysis is an expensive undertaking it is important to ensure targeted sampling approaches to maximise the DNA recovered, with clean-room laboratory techniques to minimise contamination risks, and bioinformatic methods to assess the quality of the resulting DNA.

Whereas previous Sanger sequencing could only process one sequence at a time, the development of next-generation sequencing (NGS) techniques has revolutionised ancient genetic studies, allowing millions of fragmented reads to be sequenced in parallel. As well as radically reducing the cost of generating genome-wide data, this new technology has increased the amount of data produced from a smaller amount of starting material, which is particularly important when considering rare or limited samples (Marciniak *et al.* 2015). There are two main NGS methodologies:

1) target-enriched capture arrays can be used to amplify selected loci by employing a series of baits to fish for areas of interest within the genome (Linderholm 2015); and, in comparison
2) whole genome shotgun sequencing randomly sequences all the DNA present in a sample, amplifying both the endogenous DNA from the sample in question, as well as the exogenous (contaminant) DNA that will have been co-extracted from the sample (Hofreiter *et al.* 2015).

Ancient DNA from Greece

Chilvers *et al.* (2008) suggested that DNA from the Aegean rarely survives and, of 88 skeletons studied, they managed to amplify endogenous mitochondrial DNA from only six burials (at 6.8% success rate). By contrast, Hughey *et al.* (2013) managed to generate partial ancient mtDNA from 53.6% (37 out of 69) Minoans from a cave ossuary at Haghios Charalambos, Lasithi, east-central Crete, and concluded that the Minoans were descendants of Neolithic farmers from Anatolia and the Middle East. However, it must be noted that their research was based on partial mitochondrial data. Together, these two studies highlight the importance of depositional conditions for survival of DNA, with colder, less acidic locations, such as the limestone tombs seen at the LM III Necropolis of Armenoi, being preferable.

At the time of the current analyses, there was only one Bronze Age Greek study that involved analysis of genomic DNA. Lazaridis *et al.* (2017) analysed 19 ancient individuals using capture sequencing from across the Aegean, including ten Minoans from the Final Neolithic to MM IIB cave of Hagia Charalambos, Lasithi (ca. pre-3000–1600 BC), which pre-dates the LM III Necropolis of Armenoi, and four Mycenaeans dated by the authors to 1700–1200 BC. Both groups showed ancestry similar to Neolithic farmers of the Aegean/western Anatolia and both were found to be significantly different to modern Greek populations. However, it was still possible to differentiate Minoans from Mycenaeans, with the latter exhibiting an additional source of ancestry when studied using ADMIXTURE (described in more detail below). This study also included an individual from the Armenoi Necropolis; an adult woman from Tomb 160, a single period tomb dating to Late Minoan IIIA:2 (ca. 1370–1340 BC), with an endogenous DNA content of 0.5%. She appeared noticeably different to the other Minoans in the analysis but, due to low coverage of her genome, it was not possible to conclude more about the relationships between her and the other Minoan and Mycenaean individuals.

Since completing the preparation of this Chapter, DNA data for 102 ancient individuals from Crete, the Greek mainland, and the Aegean Islands have been published (Skourtanioti *et al.* 2023). In the context of our chapter, of particular interest are the new DNA results for: 1) 28 LBA individuals from Chania in Crete; 2) 28 Early Bronze Age individuals from Hagios Charalambos in Crete; and 3) a total of 19 Late Bronze Age individuals from Tiryns, Mygdalia and Aidonia in the Peloponnese. These Bronze Age Greek populations exhibited admixed ancestry from Anatolian Neolithic, Iranian Neolithic and Western Hunter-Gatherers.

Results and discussion of our ancient DNA analyses

Endogenous DNA content

The endogenous DNA content of the 55 individuals was calculated (Table 5.1) and varied between 0.01% and 26.93%. While some tombs had consistently poor DNA survival (<0.3% in Tombs 27, 159, 189 and 206), others showed variability in DNA preservation (0.02–26.93% in Tombs 67, 149, 167, 198, 204, 208 and 210), supporting the fact that multiple factors in the burial, and storage, environments are important for ancient DNA survival (Heintzman *et al.* 2015; Kistler *et al.* 2017; Marciniak and Perry 2017). Tomb 203 showed exceptional preservation, with all eight individuals sampled having an average endogenous content of 8.3% (2.68–16.47%).

Based on the endogenous DNA contents of these 55 screened samples, 23 individuals from ten tombs were selected and sent for further in-depth NGS (Table 5.2). These 23 were compared to both modern and ancient published populations, including data from other Bronze Age Cretans (Lazaridis *et al.* 2017).

Molecular sex determination

All 23 people who were analysed using in-depth NGS were sexed genetically, with 16 females and seven males being identified (Table 5.2). These results are somewhat surprising as the sample selection was undertaken based purely upon DNA preservation. However, this sample size is small and, once we have screened the remaining skeletons and undertaken further in-depth NGS for more individuals, it is likely that this ratio will change.

Kinship analysis

In our dataset of 23 individuals, no kinship was found between people interred in different tombs. However, relationships were observed in three of the five sampled tombs with multiple individuals: Tombs 198, 203 and 210. This suggests that kinship was important to the population of the Armenoi Necropolis.

Of the three females from Tomb 198, B and Γ showed a first-degree relationship, so were either mother and daughter, or sisters. Skeleton Δ was unrelated to either of them. In Tomb 210, females E and 4 also showed a first-degree relationship to each other. As Skeleton 4 was a child, it is most likely that she was the daughter or sister of Skeleton E.

For Tomb 203, it was possible to recreate a family tree over three generations, which suggests that the idea of family was important. Seven of the eight people tested produced READ (Relationship Estimation from Ancient DNA) results suggesting either a first- or second-degree relationship. Only female ΣΤ was unrelated to any of the others from this tomb. The presence of shared uniparental markers (as described above) allowed us to reconstruct a family group (Fig. 5.4), where shared mitochondrial haplogroups represent maternal kinship and shared Y-chromosomal DNA represents paternal kinship.

It was not possible to identify evidence of a patriarchal society from the genetic analyses, but the finding of more females than males (16 and seven, respectively) Table 5.2 may indicate that LM III Crete was not a patriarchal society. However, more samples are required to explore this issue further. It is interesting to note that individuals from Hagios Charalambos analysed by Skourtanioti *et al.* (2023) showed evidence of endogamy and consanguinity, which was not present in the other Bronze Age Cretans or our results from Armenoi.

Analysis of Minoan ancestry

With the exception of Skeleton E from Tomb 149, who plots closer to Bronze Age individuals from Western Europe, the PCA plot of the other 22 individuals that we analysed from Armenoi form a homogeneous population (Fig. 5.5). They overlap both Mycenaeans from mainland Greece (ca. 1700–1200 BC) and, to a slightly lesser extent, the earlier Minoans (EM I–MM IIB, ca. 2900–1700 BC) from other parts of Crete (Lazaridis *et al.* 2017). Together these three groups fall between the earlier Greek plus Anatolian Neolithic populations (at the bottom middle of the plot) and modern Greeks (at the centre of the plot; data from Lazaridis *et al.* 2017; Stamatoyannopoulos *et al.* 2017; Mathieson *et al.* 2018; Drineas *et al.* 2019).

Despite dramatic cultural changes since the Neolithic, the population at the Late Minoan III Necropolis of Armenoi derive most of their ancestry from the Anatolian Neolithic. This ancestry was widespread in south-eastern Europe and does not necessarily reflect genetic continuity. The difference seen between the people from Armenoi and the earlier Neolithic demonstrates a small additional source of ancestry, but the origin of this source is not evident as the people from the Necropolis plot between populations related to both the Yamnaya (an Early Bronze Age culture of the Pontic-Caspian Steppe) and the Iranian Neolithic. Modern Greeks, in comparison, are shifted further towards the Yamnaya, which is evidence for significant later demographic changes. Bronze Age genomes from Crete and mainland Greece are separate from the other Bronze Age populations who originate from the Balkans (Bulgaria, Croatia and Montenegro) and Western Europe.

ADMIXTURE analysis of the Necropolis individuals comprises three main components (Fig. 5.6). Although these maximise in the Anatolian Neolithic (dark grey with zig-zag), Iranian Neolithic (mid-grey), and Western

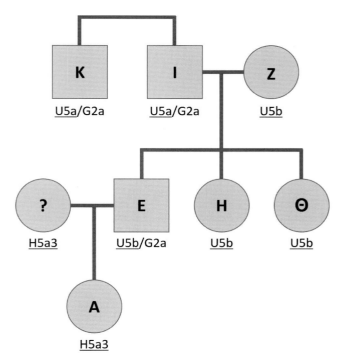

Figure 5.4. Possible kinship relationships in Tomb 203. This is one potential family tree, taking into consideration the results from READ, the genetic sex determination (squares = males; circles = females) and the uniparental data (underlined text = mitochondrial haplogroup; non-underlined text = Y-chromosomal haplogroup). Y-chromosomal haplogroup G2a has previously been found among published Minoan individuals (Lazaridis et al. *2017). The skeleton IDs are shown within each shape, with the untested individual, denoted by a question mark, included to aid interpretation of the diagram.*

Hunter-Gatherers (WHG; light grey with circles), these three groups may not have directly contributed to the Necropolis population, as the ancestry could have been incorporated into the Necropolis population via breeding with admixed groups, such as the Yamnaya. Skeleton E from Tomb 149 is, once again, clearly an outlier, whereas the rest of the Necropolis population is largely homogeneous. The previously published sample, a female from Tomb 160, is similar to the other people from the Necropolis, but she has a higher proportion of the WHG (light grey with circles) component.

As can be seen in Figure 5.6, the population from the LM III Necropolis of Armenoi derived the majority of their ancestry from the Anatolian Neolithic (dark grey with zig-zag), with additional ancestry from groups related to the Iranian Neolithic (mid-grey) or admixed groups including the Yamnaya (which would be a mixture of the mid-grey and WHG light grey with circles). This result is very similar to that found by Skourtanioti *et al.* (2023). Although this study was published after completion of this chapter, and therefore could not be included in our comparative dataset, it is in general agreement with our analysis, with the admixed ancestry of their Bronze Age Greek populations

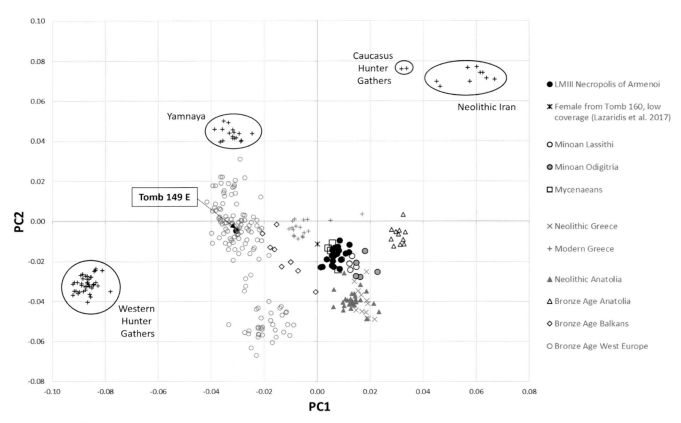

Figure 5.5. A PCA plot showing ancient populations (see key for details) projected onto a background of modern west Eurasian variation (not shown). The majority of the individuals from the LM III Necropolis of Armenoi (black dots) plot between Minoan (EM I–MM IIB ca. 2900–1700 BC) and Mycenaean (ca. 1700–1200 BC) samples (Lazaridis et al. *2017) and are distinct from earlier Neolithic and other Bronze Age populations. The outlier from Tomb 149, Skeleton E, is labelled in the plot. Of note, modern Greeks are different to Bronze Age Greek populations.*

being similar to our Armenoi individuals. As observed by Lazaridis *et al.* (2017), there is a slight difference in the ADMIXTURE proportions between Minoans and Mycenaeans, with Minoans having a higher Anatolian Neolithic component. In this regard, the people in this study appear most similar to the published Mycenaeans.

On the basis of our new data from the Necropolis, it is possible to consider one of the original aims of our study, which was to assess whether the Minoan population in Late Minoan III Crete represented:

1) genetic continuity from the Minoans;
2) an incoming Mycenaean community;
3) an admixture between the two groups; or
4) an elite takeover.

In the 'elite takeover' scenario, we would expect to see some individuals who appear to be Mycenaean-like, while others appear more Minoan-like. In this scenario a potential link between genomic differences may be related to differing tomb wealth. However, no noticeable differences can be seen between the genomic composition of the individuals tested, nor between their ancestry and the respective wealth of

their tombs as expressed in grave goods. Assessment of the PCA (Fig. 5.5) and ADMIXTURE (Fig. 5.6) plots suggests that, as already mentioned, the majority of the Necropolis individuals sampled were very similar, forming a relatively homogeneous population.

It is difficult from these analyses alone to determine with certainty if the people from the Necropolis more closely resemble earlier Minoans (EM I–MM IIB, ca. 2900–1700 BC) or Mycenaeans. However, on the PCA plot (Fig. 5.6), the people from the Necropolis have a slightly greater overlap with the Mycenaeans than the earlier EM I–MM IIB Minoans. Similarly, there are slight differences seen between the proportions of each admixture component (Fig. 5.6). The Necropolis falls between the earlier EM I–MM IIB Minoan and the later Mycenaean populations. This implies that the Necropolis population is not a complete replacement but nor is it a simple genetic continuation. Instead, our results suggest the possibility that the Necropolis population are a result of one or many admixture events between Minoan and Mycenaean populations. However, it should be noted that further statistical testing is necessary to fully investigate the genomic ancestry at the Armenoi, and that our current research is

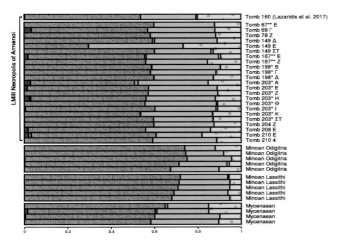

Figure 5.6. ADMIXTURE results from the LM III Necropolis of Armenoi compared to published data from earlier Bronze Age Crete and later Bronze Age Mycenae on mainland Greece (Lazaridis et al. 2017). The three main components shown are: Anatolian Neolithic (dark grey with zig-zag), Iranian Neolithic (mid-grey), and Western Hunter-Gatherers (light grey with circles). Tomb 149 E is, once again, clearly an outlier, but otherwise the Armenoi population is largely homogeneous. No other significant ancestry was identified amongst the Bronze Age Greek samples, and the presence of other components are most likely statistical noise caused by the low genomic coverage of the ancient DNA.

influenced by the availability of reference samples. In light of the new results published recently by Skourtanioti et al. (2023), we are currently undertaking a more nuanced analysis of our Armenoi individuals for publication in the scientific literature (Foody et al. in prep.).

Finally, the ancestry results clearly show that the male Skeleton E from Tomb 149 consistently appears as an outlier, with his genome resembling Bronze Age populations from Western Europe (Fig. 5.5). Therefore, he was most likely an immigrant to the area.

Radiocarbon dating

Of the eight samples sent for dating, six returned radiocarbon dates. As can be seen in Figure 5.7, they exhibit rather large ranges (between 150 years for Tomb 149 ΣT and 265 years for Tomb 149 Δ), which suggests that we will not be able to gain any nuanced picture of tomb use via radiocarbon dating of individuals. All the dates overlap with each other, largely due to the shape of the radiocarbon calibration curve at this time period, which exhibits a sharp but transient reversal at approximately 1330 cal BC. This has the effect of broadening the calibrated age range for the determinations into multimodal probability density curves (Table 5.3; Fig. 5.7). This is also seen in the OxCal R_combine test for the six individual dates, where a clear bimodal distribution can be seen, with calibrated age ranges of 1421–1373 and

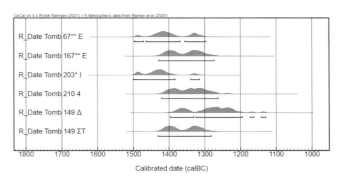

Figure 5.7. Radiocarbon results from the six individuals dated successfully from across the LMIII Necropolis of Armenoi, calibrated to 2σ using OxCal (Bronk Ramsey 2009) and the most recent calibration curve, IntCal20 (Reimer et al. 2020). The tombs have been ordered based on their location within the Necropolis, from north to south, as shown in Figure 5.1.

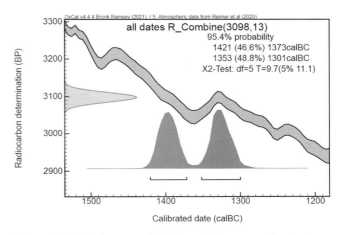

Figure 5.8. OxCal output of the R_combine test on the six dates. These combined with a chi-squared score of 9.7, which fails the 5% confidence limit score of 3.8, highlighting that the six dates cannot be encapsulated by a single date range. Instead, the dates are bi-modal, with ranges of 1421–1373 and 1353–1301 cal BC at a 95.4% confidence interval.

1353–1301 cal BC (Fig. 5.8). However, overall, the calibrated dates broadly fit the typology contextual dates for the Necropolis of 1390–1190 BC.

While there was enough collagen to undertake dietary stable isotopic analysis on Skelton Γ from Tomb 69 (Table 5.1), its yield was not sufficient for dating purposes. The other sample to fail with conventional dating methods was the outlier individual from Tomb 149 (Skeleton E; Fig. 5.5). As we wanted to see if he was contemporaneous with the local individuals (Δ and ΣT) buried alongside, this sample was sent to ORAU for more detailed analysis, but again, no radiocarbon date could be obtained. This absence of extractable collagen from Skeleton E was unsurprising given that stable isotopic analysis indicated a C:N ratio of 4.9, well above

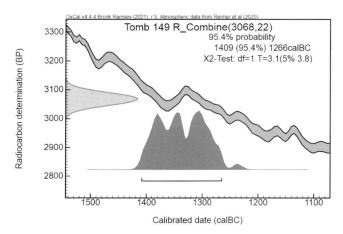

Figure 5.9. OxCal output of the R_combine test on the two individuals dated from Tomb 149. These combined successfully to give a calibrated date of 1409–1266 cal BC at a 95.4% confidence interval.

the range of values expected from diagenetically unaltered collagen (as discussed above). However, the endogenous DNA content from this sample was 3.7% (Table 5.2) and gave useable genetic data for downstream NGS analyses. The absence of any collagen, therefore, is further proof that, although there does appear to be a general link between preservation of collagen and survival of DNA, this relationship is not always linear, with many factors being involved (Campos *et al.* 2012). The dates resulting from the other two individuals from Tomb 149 (Δ and ΣT) are shown in Table 5.3. These were analysed using an OxCal R_combine test to test whether they were similar to each other, and the results (a chi-squared score of 3.1) passed the 5% confidence limit score of 3.8, suggesting that Tomb 149 was in use during the period covered by the combined calibrated date range of 1409–1266 cal BC at 95.4% confidence (Fig. 5.9).

Conclusions

- As already determined by Richards and Hedges in their 2008 study, the diet of the population at the LM III Necropolis of Armenoi is terrestrial based and appears to have been relatively rich in animal protein but with no significant input of marine fish into the diet. There are no differentiations based on sex or status.
- When the genome data are compared with the earlier Minoan populations from eastern Crete, while the Necropolis does not appear to be a simple continuation, neither is there evidence of complete replacement by mainland Mycenaeans or an elite takeover. Instead, using the results presented here, the hypothesis that admixture occurred between Minoans and the incoming Mycenaeans is plausible. However, further statistical analysis is required to produce more conclusive results.
- The presence of a genetic outlier in Tomb 149 shows both that migrants were present in Late Minoan Crete,

and that they were not treated any differently in death from the local people, being buried in a Necropolis tomb alongside non-related local individuals.
- While no kinship was found between people interred in different tombs, two tombs (Tomb 198 and Tomb 210) each contained a mother–daughter or sister relationship. Tomb 203 contained a multi-generational family group which is indicative that tombs were used as family burial places and that kinship was important to the people who lived in the 'city' that built the Necropolis.

Prior to the publication by Skourtanioti *et al.* (2023), this study was the first large scale genomic sampling from the Greek Bronze Age. However, it is still one of the first instances where a multi-disciplinary approach has been conducted on a large, well curated necropolis assemblage. This, coupled with the well documented archaeological investigation of the site, has provided us with a unique opportunity to answer localised archaeological questions, such as kinship and diet, as well as larger stories about migration and culture change, and our goal is to build upon this work with future study of the people from the LM III Necropolis of Armenoi.

Appendix: species identification of 13 animal bones from the Necropolis

Zooarchaeological assessment

Prior to analysis, the 13 animal bones collected from the Necropolis were assessed morphologically to identify each to the species level. This was undertaken by Angelos Hadjikoumis and Umberto Albarella of the University of Sheffield, and results are shown in Table 5.5. The long bone in Tomb 146 (ARM01) was identified as a fully fused sheep bone, which was at least 2.5 years old at death (Zeder 2006). The 12 bones in Tomb 159 were identified as either sheep/goat or hare/rabbit.

It is often difficult to distinguish between sheep and goat bones in the archaeological record using osteological methods alone, especially in skeletal elements that are unfused (Buckley *et al.* 2010). Of the nine sheep/goat samples, six could not be assigned unambiguously to sheep or to goat. A fused distal metatarsal (ARM03), with an age at death of at least 18 months, was identified as a probable goat; however, this was questionable due to damage to the bone. A fusing distal humerus (ARM08), with an age at death of 6–18 months, was identified as a sheep. The mandible (ARM13) was identified as a sheep due to the characteristics of the third deciduous premolar and the eruption pattern suggests that it was 12–24 months old.

The four hare/rabbit bones were a femur (ARM04) and three humeri (ARM09–ARM11). Due to their large size, and from knowledge of historic rabbit-hare habitat ranges (Irving-Pease *et al.* 2018), the bones were assessed to the genus level as being *Lepus* – most likely the European hare, *Lepus europaeus* – rather than rabbit (*Oryctolagus*). The hare bones

Table 5.5. Sample details for the 13 animal bones collected from Tombs 146 and 159. Results from both the zooarchaeological and ZooMS analyses are compared.

Tomb	Code	Element sampled	Age	Fusion	Completeness	Zooarch ID	ZooMS ID	Species
146	ARM01	femur	>2.5 years	fully fused	complete	*Ovis* sp.	*Ovis* sp.	sheep
	ARM02	metatarsal	–	unfused	distal end	ovicaprid	Ovis sp.	sheep
	ARM03	metatarsal	>1.5 years	fused	distal end	*Capra* sp.?	*Capra* sp.	goat
	ARM04	femur	–	fused	proximal end and shaft	*Lepus* sp.	*Lepus* sp.	hare
	ARM05	humerus	–	unfused/fusing	partial proximal end and shaft	ovicaprid	*Ovis* sp.	sheep
	ARM06	femur	–	unfused	proximal end	ovicaprid	*Ovis* sp.	sheep
159	ARM07	femur	–	unfused	proximal end	ovicaprid	*Ovis* sp.	sheep
	ARM08	humerus	<1.5 years	fusing	distal end	*Ovis* sp.	*Ovis* sp.	sheep
	ARM09	humerus	–	fused	distal end	*Lepus* sp.	lagomorph	hare
	ARM10	humerus	–	fused	distal end and shaft	*Lepus* sp.	lagomorph	hare
	ARM11	humerus	–	fused	distal end	*Lepus* sp.	lagomorph	hare
	ARM12	tibia	–	unfused	distal end and shaft	ovicaprid	*Ovis* sp.	sheep
	ARM13	mandible	1–2 years	n/a	dp3, dp4, 1st molar present	*Ovis* sp.	*Ovis* sp.	sheep

were noted as being large for southern European hare. As there was no evidence of human alteration, such as butchery, cooking or gnawing, these hares appear not to have been consumed but were more likely intrusive in the burial environment.

From the 12 bones sampled from Tomb 159, the minimum number of individuals (MNI) for hares was two and for sheep/goat was three. No cut marks were found on any of the bones. Gnawing was found on the potential goat metatarsal (ARM03), and the puncture marks suggested that this was from a medium-sized mammal.

ZooMS analysis

Zooarchaeological Mass Spectrometry (ZooMS) is a method of identifying animals to the taxa, or even genus level, targeting the peptides found in Type 1 collagen. This is the most abundant protein in vertebrates, surviving particularly well in archaeological bone, and usually for much longer time periods than DNA (Buckley *et al.* 2010). It has a triple helical structure, comprising of two identical alpha 1 (α1) chains and a genetically distinct alpha 2 (α2) chain (Buckley 2018).

The analytical tool used for ZooMS is Matrix-Assisted Laser Desorption/Ionization Time Of Flight (MALDI-TOF) Mass Spectrometry. This method irradiates the sample with a laser, and the time of flight of the peptide ions released is measured over a known distance, with larger peptide ions taking longer to traverse the distance than smaller ones. This produces a spectra that displays the relative abundance of the ions detected, and creates a peptide mass fingerprint, which can then be compared to known spectra in order to identify the taxa or genus (Buckley *et al.* 2017). This method is of particular use when differentiating sheep from goats. Both species have long been major constituents of livestock populations but, as mentioned above, are very difficult to differentiate through zooarchaeological assessment of bone alone (Buckley *et al.* 2010).

Analyses were performed by Michael Buckley at the University of Manchester, following the protocol outlined in Buckley *et al.* (2010) and the resulting spectra were compared to a known database of samples held at the Manchester Institute of Biotechnology. The results of ZooMS were able to identify four samples as lagomorphs (hare or rabbit), and nine as ovicaprids (sheep or goat). Of the lagomorphs, ARM04 could be further identified as a probable *Lepus* (hare) through presence of the diagnostic peptide marker 2808. One of the ovicaprids, ARM03, had the characteristic peptide marker 3094 unique to goat (*Capra*), while the other eight were identified as sheep (*Ovis*). These results support the zoological assignment, and comparisons between the zooarchaeological and ZooMS analyses are shown in Table 5.5.

Acknowledgements

We would like to thank Drs Yannis Tzedakis and Holley Martlew, and Professor Robert Arnott, without whom a genetic study of the inhabitants of the chamber tombs in the Armenoi Necropolis would not have been possible. They facilitated access to, and selection of, the samples and provided background information on the site. We would also like to thank Professor Michael S. Tite, Scientific Editor for this volume, for his input on earlier drafts of the Chapter, as well as to Dr Valeria Mattiangeli (Trinity College Dublin) for screening of the human samples for endogenous DNA content; Steve Litherland and Philip Mann for reproduction of the site map; Professor Umberto Albarella and Dr Angelos Hadjikoumis (University of Sheffield) for zooarchaeological assessment;

Dr Mike Buckley (University of Manchester) for help with the ZooMS analysis; and The Leverhulme Trust and the University of Huddersfield URF for funding the research.

Bibliography

Alexander, D.H., Novembre, J. and Lange, K. (2009) Fast model-based estimation of ancestry in unrelated individuals. *Genome Research* 19, 1655–64.

Ambrose, S.H. (1990) Preparation and characterization of bone and tooth collagen for isotopic analysis. *Journal of Archaeological Science* 17, 431–51.

Bogaard, A., Heaton, T.H.E., Poulton, P. and Merbach, I. (2007) The impact of manuring on nitrogen isotope ratios in cereals: archaeological implications for reconstruction of diet and crop management practices. *Journal of Archaeological Science* 34, 335–43.

Brock, F., Higham, T., Ditchfield, P. and Bronk Ramsey, C. (2010) Current pretreatment methods for AMS radiocarbon dating at the Oxford Radiocarbon Accelerator Unit (ORAU). *Radiocarbon* 52, 103–12.

Bronk Ramsey, C. (2009) Bayesian analysis of radiocarbon dates. *Radiocarbon* 51, 337–60.

Buckley, M. (2018) Zooarchaeology by mass spectrometry (ZooMS) collagen fingerprinting for the species identification of archaeological bone fragments. In C.M. Giovas and M.J. LeFebvre (eds), *Zooarchaeology in Practice: case studies in methodology and interpretation in archaeofaunal analysis*, 227–47. New York, Springer.

Buckley, M., Harvey, V.L. and Chamberlain, A.T. (2017) Species identification and decay assessment of Late Pleistocene fragmentary vertebrate remains from Pin Hole Cave (Creswell Crags, UK) using collagen fingerprinting. *Boreas* 46, 402–11. [https://doi.org/10.1111/bor.12225]

Buckley, M., Kansa, S.W., Howard, S., Campbell, S., Thomas-Oates, J. and Collins, M. (2010) Distinguishing between archaeological sheep and goat bones using a single collagen peptide. *Journal of Archaeological Science* 37, 13–20.

Campos, P.F., Craig, O.E., Turner-Walker, G., Peacock, E., Willerslev, E. and Gilbert, M.T.P. (2012) DNA in ancient bone – where is it located and how should we extract it? *Annals of Anatomy* 194, 7–16.

Chilvers, E.R., Bouwman, A.S., Brown, K.A., Arnott, R.G., Prag, J.N.W. and Brown, T.A. (2008) Ancient DNA in human bones from Neolithic and Bronze Age sites in Greece and Crete. *Journal of Archaeological Science* 35, 2707–14.

Coplen, T.B., Brand, W.A., Gehre, M., Gröning, M., Meijer, H.A.J., Toman, B. and Verkouteren, R.M. (2006) New guidelines for δ13C measurements. *Analytical Chemistry* 78, 2439–41.

DeNiro, M.J. (1985) Postmortem preservation and alteration of *in vivo* bone collagen isotope ratios in relation to palaeodietary reconstruction. *Nature* 317, 806–9.

Dotsika, E., Diamantopoulos, G., Lykoudis, S., Gougoura, S., Kranioti, E., Karalis, P., Michael, D., Samartzidou, E. and Palaigeorgiou, E. (2019) Establishment of a Greek food database for palaeodiet reconstruction: case study of human and fauna remains from Neolithic to Late Bronze Age from Greece. *Geosciences* 9(4), 165. [https://doi.org/10.3390/geosciences9040165]

Drineas, P., Tsetsos, F., Plantinga, A., Lazaridis, I., Yannaki, E., Razou, A., Kanaki, K., Michalodimitrakis, M., Perez-Jimenez, F., de Silvestro, G., Renda, M.C., Stamatoyannopoulos, J.A., Kidd, K.K., Browning, B.L., Paschou, P. and Stamatoyannopoulos, G. (2019) Genetic history of the population of Crete. *Annals of Human Genetics* 83, 373–88. [https://doi.org/10.1111/ahg.12328]

Gamba, C., Jones, E.R., Teasdale, M.D., McLaughlin, R.L., Gonzalez-Fortes, G., Mattiangeli, V., Domboróczki, L., Kővári, I., Pap, I., Anders, A., Whittle, A., Dani, J., Raczky, P., Higham, T.F.G., Hofreiter, M., Bradley, D.G. and Pinhasi, R. (2014) Genome flux and stasis in a five millennium transect of European prehistory. *Nature Communications* 5, 5257. [https://doi.org/10.1038/ncomms6257]

Hallager, E. (2010) Late Bronze Age: Crete. In E.H. Cline (ed.), *The Oxford Handbook of the Bronze Age Aegean*, 149–59. Oxford, Oxford University Press.

Hansen, H.B., Damgaard, P.D., Margaryan, A., Stenderup, J., Lynnerup, N., Willerslev, E. and Allentoft, M.E. (2017) Comparing ancient DNA preservation in petrous bone and tooth cementum. *PLoS One* 12, e0170940. [https://doi.org/10.1371/journal.pone.0170940]

Heintzman, P.D, Soares, A.E.R., Chang, D. and Shapiro, B. (2015) Paleogenomics. *Reviews in Cell Biology and Molecular Medicine* 1, 243–67. [https://pgl.soe.ucsc.edu/mcb.201500020.pdf]

Herrera, R.J. and Garcia-Bertrand, R. (2018) *Ancestral DNA, Human Origins, and Migrations*. Cambridge MA, Academic Press.

Hofreiter, M., Paijmans, J.L., Goodchild, H., Speller, C.F., Barlow, A., Fortes, G.G., Thomas, J.A., Ludwig, A. and Collins, M.J. (2015) The future of ancient DNA: Technical advances and conceptual shifts. *BioEssays* 37, 284–93. [https://doi.org:10.1002/bies.201400160]

Hughey, J.R., Paschou, P., Drineas, P., Mastropaolo, D., Lotakis, D.M., Navas, P.A., Michalodimitrakis, M., Stamatoyannopoulos, J.A. and Stamatoyannopoulos, G. (2013) A European population in Minoan Bronze Age Crete. *Nature Communications* 4, 1861. [https://doi.org/10.1038/ncomms2871]

Irving-Pease, E.K., Frantz, L.A., Sykes, N., Callou, C. and Larson, G. (2018) Rabbits and the specious origins of domestication. *Trends in Ecology and Evolution* 33, 49–152. [https://doi.org/10.1016/j.tree.2017.12.009]

Jobling, M.A., Hollox, E., Hurles, M., Kivisild, T. and Tyler-Smith, C. (2013) *Human Evolutionary Genetics* (2nd edn). New York, Garland Science.

Kistler, L., Ware, R., Smith, O., Collins, M. and Allaby, R.G. (2017) A new model for ancient DNA decay based on paleogenomic meta-analysis. *Nucleic Acids Research* 45, 6310–20. [https://doi.org/10.1093/nar/gkx361]

Kuhn, J.M.M., Jakobsson, M. and Günther, T. (2018) Estimating genetic kin relationships in prehistoric populations. *PLoS One* 13, e0195491. [https://doi.org/10.1371/journal.pone.0195491]

Lamb, A.L, Evans, J.E., Buckley, R. and Appleby, J. (2014) Multi-isotope analysis demonstrates significant lifestyle changes in King Richard III. *Journal of Archaeological Science* 50, 559–65.

Lazaridis, I., Patterson, N., Mittnik, A., Renaud, G., Mallick, S., Kirsanow, K., Sudmant, P., Schraiber, J., Castellano, S., Lipson, M., Berger, B., Economou, C., Bollongino, R., Fu, Q., Bos, K., Nordenfelt, S., Li, H., de Filippo, C., Prüfer, K., Sawyer, S., Posth, C., Haak, W., Hallgren, F., Fornander, E., Rohland,

N., Delsate, D., Francken, M., Guinet, J.-M., Wahl, J., Ayodo, G., Babiker, H., Bailliet, G., Balanovska, E., Balanovsky, O., Barrantes, R., Bedoya, G., Ben-Ami, H., Bene, J., Berrada, F., Bravi, C., Brisighelli, F., Busby, G., Cali, F., Churnosov, M., Cole, D., Corach, D., Damba, L., van Driem, G., Dryomov, S., Dugoujon, J.-M., Fedorova, S., Gallego Romero, I., Gubina, M., Hammer, M., Henn, B., Hervig, T., Hodoglugil, U., Jha, A., Karachanak-Yankova, S., Khusainova, R., Khusnutdinova, E., Kittles, R., Kivisild, T., Klitz, W., Kucinskas, V., Kushniarevich, A., Laredj, L., Litvinov, S., Loukidis, T., Mahley, R., Melegh, B., Metspalu, E., Molina, J., Mountain, J., Näkkäläjärvi, K., Nesheva, D., Nyambo, T., Osipova, L., Parik, J., Platonov, F., Posukh, O., Romano, V., Rothhammer, F., Rudan, I., Ruizbakiev, R., Sahakyan, H., Sajantila, A., Salas, A., Starikovskaya, E., Tarekegn, A., Toncheva, D., Turdikulova, S., Uktveryte, I., Utevska, O., Vasquez, R., Villena, M., Voevoda, M., Winkler, C., Yepiskoposyan, L., Zalloua P., Zemunik T., Cooper A., Capelli C., Thomas M., Ruiz-Linares A., Tishkoff, S., Singh, L., Thangaraj, K., Villems, R., Comas, D., Sukernik, R., Metspalu, M., Meyer, M., Eichler, E., Burger, J., Slatkin, M., Pääbo, S., Kelso, J., Reich, D. and Krause, J. (2014) Ancient human genomes suggest three ancestral populations for present-day Europeans. *Nature* 513, 409–13. [https://doi.org/10.1038/nature13673]

Lazaridis, I., Nadel, D., Rollefson, G., Merrett, D., Rohland, N., Mallick, S., Fernandes, D., Novak, M., Gamarra, B., Sirak, K., Connell, S., Stewardson, K., Harney, E., Fu, Q., Gonzalez-Fortes, G., Jones, E., Roodenberg, S., Lengyel, G., Bocquentin, B., Monge, J., Gregg, M., Eshed, V., Mizrahi, A.-S., Meiklejohn, C., Gerritsen, F., Bejenaru, L., Blüher, M., Campbell, A., Cavalleri, G., Comas, D., Froguel, P., Gilbert, E., Kerr, S., Kovacs, P., Krause, J., McGettigan, D., Merrigan, M., Merriwether, D., O'Reilly, S., Richards, M., Semino, O., Shamoon-Pour, M., Stefanescu, G., Stumvoll, M., Tönjes, A., Torroni, A., Wilson, J., Yengo, L., Hovhannisyan, N., Patterson, N., Pinhasi, R. and Reich, D. (2016) Genomic insights into the origin of farming in the ancient Near East. *Nature* 536, 419–24. [https://doi.org/10.1038/nature19310]

Lazaridis, I., Mittnik, A., Patterson, N., Mallick, S., Rohland, N., Pfrengle, S., Furtwängler, A., Peltzer, A., Posth, C., Vasilakis, A., McGeorge, P.J.P., Konsolaki-Yannopoulou, E., Korres, G., Martlew, H., Michalodimitrakis, M., Özsait, M., Özsait, N., Papathanasiou, A., Richards, M., Roodenberg, S.A., Tzedakis, Y., Arnott, R., Fernandes, D.M., Hughey, J.R., Lotakis, D.M., Navas, P.A., Maniatis, Y., Stamatoyannopoulos, J.A., Stewardson, K., Stockhammer, P., Pinhasi, R., Reich, D., Krause, J. and Stamatoyannopoulos G. (2017) Genetic origins of the Minoans and Mycenaeans. *Nature* 548, 214–18. [https://doi.org/10.1038/nature23310]

Linderholm, A. (2015) Ancient DNA: the next generation–chapter and verse. *Biological Journal of the Linnean Society* 117, 150–60. [https://doi.org/10.1111/bij.12616]

Longin, R. (1971) New method of collagen extraction for radiocarbon dating. *Nature* 230, 241–2. [https://doi.org/10.1038/230241a0]

MacHugh, D.E., Edwards, C.J., Bailey, J.F., Bancroft, D.R. and Bradley, D.G. (2000) The extraction and analysis of ancient DNA from bone and teeth: a survey of current methodologies. *Ancient Biomolecules* 3, 81–102.

Marciniak, S. and Perry, G.H. (2017) Harnessing ancient genomes to study the history of human adaptation. *Nature Reviews Genetics* 18, 659–74. [https://doi.org/10.1038/nrg.2017.65]

Marciniak, S., Klunk, J., Devault, A., Enk, J. and Poinar, H.N. (2015) Ancient human genomics: the methodology behind reconstructing evolutionary pathways. *Journal of Human Evolution* 79, 21–34. [https://doi.org/10.1016/j.jhevol.2014.11.003]

Martiniano, R., Coelho, C., Ferreira, M.T., Neves, M.J., Pinhasi, R. and Bradley, D.G. (2014) Genetic evidence of African slavery at the beginning of the trans-Atlantic slave trade. *Scientific Reports* 4, 5994. [https://doi.org/10.1038/srep05994]

Mathieson, I., Alpaslan-Roodenberg, S., Posth, C., Szécsényi-Nagy, A., Rohland, N., Mallick, S., Olalde, I., Broomandkhoshbacht, N., Candilio, F., Cheronet, O., Fernandes, D., Ferry, M., Gamarra, B., Fortes, G.G., Haak, W., Harney, E., Jones, E., Keating, D., Krause-Kyora, B., Kucukkalipci, I., Michel, M., Mittnik, A., Nägele, K., Novak, M., Oppenheimer, J., Patterson, N., Pfrengle, S., Sirak, K., Stewardson, K., Vai, S., Alexandrov, S., Alt, K.W., Andreescu, R., Antonović, D., Ash, A., Atanassova, N., Bacvarov, K., Balázs Gusztáv, M.B., Bocherens, H., Bolus, M., Boroneanţ, A Boyadzhiev, Y., Budnik, A., Burmaz, J., Chohadzhiev, S., Conard, N.J., Cottiaux, R., Čuka, M., Cupillard, C., Drucker, D.G., Elenski, N., Francken, M., Galabova, B., Ganetsovski, G., Gély, B., Hajdu, T., Handzhyiska, V., Harvati, K., Higham, T., Iliev, S., Janković, I., Karavanić, I., Kennett, D.J., Komšo, D., Kozak, A., Labuda, D., Lari, M., Lazar, C., Leppek, M., Leshtakov, K., Vetro, D.L., Los, D., Lozanov, I., Malina, M., Martini, F., McSweeney, K., Meller, H., Menđušić, M., Mirea, P., Moiseyev, V., Petrova, V., Price, T.D., Simalcsik, A., Sineo, L., Šlaus, M., Slavchev, V., Stanev, P., Starović, A., Szeniczey, T., Talamo, S., Teschler-Nicola, M., Thevenet, C., Valchev, I., Valentin, F., Vasilyev, S., Veljanovska, F., Venelinova, S., Veselovskaya, E., Viola, B., Virag, C., Zaninović, J., Zäuner, S., Stockhammer, P.W., Catalano, G., Krauß, R., Caramelli, D., Zariņa, G., Gaydarska, B., Lillie, M., Nikitin, A.G., Potekhina, I., Papathanasiou, A., Borić, D., Bonsall, C., Krause, J., Pinhasi, R. and Reich, D. (2018) The genomic history of southeastern Europe. *Nature* 555, 197–203. [https://doi.org/10.1038/nature25778]

Mays, S. (2010) *The Archaeology of Human Bones*. London, Routledge.

Meyer, M. and Kircher, M. (2010) Illumina sequencing library preparation for highly multiplexed target capture and sequencing. *Cold Spring Harbor Protocols* 6, 5448.

Nafplioti, A. (2016) Eating in prosperity: first stable isotope evidence of diet from Palatial Knossos. *Journal of Archaeological Science: Reports* 6, 42–52. [https://doi.org/10.1016/j.jasrep.2016.01.017]

Nitsch, E., Andreou, S., Creuzieux, A., Gardeisen, A., Halstead, P., Isaakidou, V., Karathanou, A., Kotsachristou, D., Nikolaidou, D., Papanthimou, A., Petridou, C., Triantaphyllou, S., Valamoti, S.M., Vasileiadou, A. and Bogaard, A. (2017) A bottom-up view of food surplus: using stable carbon and nitrogen isotope analysis to investigate agricultural strategies and diet at Bronze Age Archontiko and Thessaloniki Toumba, northern Greece. *World Archaeology* 49, 105–37.

Papathanasiou, A. (2015) Stable isotope analyses in Neolithic and Bronze Age Greece: an overview. *Hesperia Supplements* 49, 25–55.

Papathanasiou, A. and Richards, M.P. (2015) Summary: patterns in the carbon and nitrogen isotope data through time. *Hesperia Supplements* 49, 195–203.

Patterson, N., Price, A. and Reich, D. (2006) Population structure and eigenanalysis. *PLoS Genetics* 2, e190. [https://doi.org/10.1371/journal.pgen.0020190]

Patterson, N., Moorjani, P., Luo, Y., Mallick, S., Rohland, N., Zhan, Y., Genschoreck, T., Webster, T. and Reich, D. (2012) Ancient admixture in human history. *Genetics* 192(3), 1065–93. [https://doi.org/10.1534/genetics.112.145037]

Pinhasi, R., Fernandes, D., Sirak, K., Novak, M., Connell, S., Alpaslan-Roodenberg, S., Gerritsen, F., Moiseyev, V., Gromov, A., Raczky, P., Anders, A., Pietrusewsky, M., Rollefson, G., Jovanovic, M., Trinhhoang, H., Bar-Oz, G., Oxenham, M., Matsumura, H. and Hofreiter, M. (2015) Optimal ancient DNA yields from the inner ear part of the human petrous bone. *PLoS One* 10, e0129102. [https://doi.org/10.1371/journal.pone.0129102]

Reimer, P.J., Austin, W.E.N., Bard, E., Bayliss, A., Blackwell, P.G., Bronk Ramsey, C., Butzin, M., Cheng, H., Edwards, R.L., Friedrich, M., Grootes, P.M., Guilderson, T.P., Hajdas, I., Heaton, T.J., Hogg, A.G., Hughen, K.A., Kromer, B., Manning, S.W., Muscheler, R., Palmer, J.G., Pearson, C., van der Plicht, J., Reimer, R.W., Richards, D.A., Scott, E.M., Southon, J.R., Turney, C.S.M., Wacker, L., Adolphi, F., Büntgen, U., Capano, M., Fahrni, S.M., Fogtmann-Schulz, A., Friedrich, R., Köhler, P., Kudsk, S., Miyake, F., Olsen, J., Reinig, F., Sakamoto, M., Sookdeo, A. and Talamo, S. (2020) The IntCal20 Northern Hemisphere Radiocarbon Age Calibration Curve (0–55 Cal KBP). *Radiocarbon* 62, 725–57. [https://doi.org/10.1017/RDC.2020.41]

Richards, M.P. (2015) Stable isotope analysis of bone and teeth as a means for reconstructing past human diets in Greece. *Hesperia Supplements* 49, 15–23.

Richards, M.P. and Hedges, R.E.M. (2008) Stable isotope results from the sites of Gerani, Armenoi and Mycenae. In Y. Tzedakis, H. Martlew and M.K. Jones (eds), *Archaeology Meets Science: biomolecular and site investigations in Bronze Age Greece*, 220–30. Oxford, Oxbow Books.

Schoeninger, M.J. and Moore, K. (1992) Bone stable isotope studies in archaeology. *Journal of World Prehistory* 6, 247–96.

Schulting, R.J. (1998) Slighting the sea: stable isotope evidence for the transition to farming in northwestern Europe. *Documenta Praehistorica* 25, 18.

Skoglund, P., Storå, J., Götherström, A. and Jakobsson, M. (2013) Accurate sex identification of ancient human remains using DNA shotgun sequencing. *Journal of Archaeological Science* 40(12), 4477–82. [https://doi.org/10.1016/j.jas.2013.07.004]

Skourtanioti, E., Ringbauer, H., Gnecchi Ruscone, G.A., Bianco, R.A., Burri, M., Freund, C., Furtwängler, A., Gomes Martins, N.F., Knolle, F., Neumann, G.U., Tiliakou, A., Agelarakis, A., Andreadaki-Vlazaki, M., Betancourt, P., Hallager, B.P., Jones, O.A., Kakavogianni, O., Kanta, A., Karkanas, P., Kataki, E., Kissas, K., Koehl, R., Kvapil, L., Maran, J., McGeorge, P.J.P., Papadimitriou, A., Papathanasiou, A., Papazoglou-Manioudaki, L., Paschalidis, K., Polychronakou-Sgouritsa, N., Preve, S., Prevedorou, E.-A., Price, G., Protopapadaki, E., Schmidt-Schultz, T., Schultz, M., Shelton, K., Wiener, M.H., Krause, J., Jeong, C. and Stockhammer, P.W. (2023) Ancient DNA reveals admixture history and endogamy in the prehistoric Aegean. *Nature Ecology and Evolution* 7, 290–303. [https://doi.org/10.1038/s41559-022-01952-3]

Stamatoyannopoulos, G., Bose, A., Teodosiadis, A., Tsetsos, F., Plantinga, A., Psatha, N., Zogas, N., Yannaki, E., Zalloua, P., Kidd, K.K., Browning, B.L., Stamatoyannopoulos, J., Paschou, P. and Drineas, P. (2017) Genetics of the Peloponnesian populations and the theory of extinction of the medieval Peloponnesian Greeks. *European Journal of Human Genetics* 25, 637–45. [https://doi.org/10.1038/ejhg.2017.18]

Triantaphyllou, S., Richards, M.P., Zerner, C. and Voutsaki, S. (2008) Isotopic dietary reconstruction of humans from Middle Bronze age Lerna, Argolid, Greece. *Journal of Archaeological Science* 35, 3028–34. [https://doi.org/10.1016/j.jas.2008.06.018]

Twiss, K. (2012) The archaeology of food and social diversity. *Journal of Archaeological Research* 20, 357–95.

Tzedakis, Y. and Kolivaki, V. (2018) Background and history of the excavation. In Y. Tzedakis, H. Martlew and R. Arnott (eds) *The Late Minoan III Necropolis of Armenoi* I, 1–18. Philadelphia PA, INSTAP Academic Press.

van Klinken, G.J (1999) Bone collagen quality indicators for palaeodietary and radiocarbon measurements. *Journal of Archaeological Science* 26, 687–95. [https://doi.org/10.1006/jasc.1998.0385]

Yang, D.Y., Eng, B., Waye, J.S., Dudar, J.C. and Saunders, S.R. (1998) Technical note: improved DNA extraction from ancient bones using silica-based spin columns. *American Journal of Physical Anthropology* 105, 539–43.

Zeder, M.A. (2006) Reconciling rates of long bone fusion and tooth eruption and wear in sheep (*Ovis*) and goat (*Capra*). In D. Ruscillo (ed.) *Recent Advances in Ageing and Sexing Animal Bones*, 87–118. Oxford, Oxbow Books.

6

The human dimension: the archaeological significance of the scientific results

Holley Martlew and Yannis Tzedakis

Introduction

The intention of this chapter is to shine a light on the lives of those who came before us. The summary of analyses presented in this chapter were performed on skeletal and faunal material as reported in detail in Chapters 3–5. The human skeletal material which produced the most outstanding scientific results came from Tombs 55, 89, 149, 159, 198, 203 and 210. The object of the chapter is to add a human dimension, to show how the scientific results, ancient DNA, stable isotope analysis, osteological analysis, and radiocarbon dating, add to our understanding of Minoan society in the Late Bronze Age, as seen through the microcosm of a great Necropolis and the 'city' that built it. To achieve this, the chapter summarises the results and gives background information on the tombs, one or more of whose inhabitants yielded successful results, to show how they open a window into Minoan society.

The Total Station survey

A key component of the site recording was the production of accurate Tomb Plan Pages (*e.g.* Fig. 6.1). To be able to realise the Tomb Plan Pages, a survey of the Necropolis was carried out over a period of years, directed by the late Stephen Litherland and supported by Philip Mann and a group of assistants, using a Total Station Theodolite (TST). This combined horizontal and vertical angle measurements with an infra-red distance measurement unit (EDM). Measurements were recorded digitally and processed using survey software to create a series of 3-D Cartesian coordinates relative to the instrument position, from which a map or plan or elevation can be derived.

The object of the Total Station survey of the Necropolis was to produce an accurate plan for each chamber tomb and built structure. This made it possible for a plan and ultimately, a Tomb Plan Page, to be created for each tomb,

by Magdalena Wachnik, the Armenoi Project's archaeological illustrator.

Tomb Plan Pages

A Tomb Plan Page is provided in the discussion of each of the tombs discussed in this chapter, and in Chapter 9. A Tomb Plan Page includes: a plan of the tomb based on the Total Station Survey, chronology, dimensions, type of entrance, MNI (minimum number of skeletons), year of excavation, year of survey, map of the Necropolis, placement in the Necropolis, and a photograph. This is to give the reader the most comprehensive background possible. These pages are an essential part of the 'encyclopaedia' of the Necropolis that will form Volume III (see Chapter 1).

Catalogue entries and archives

Abbreviated catalogue entries for each tomb discussed in this Chapter and in Chapter 9 are included at the end of the respective chapters. Vicky Kolivaki was responsible for producing the catalogue entries and archives, assisted in the latter by Michael Jones.

Excavation notebooks

The notes and drawing made at the time of a tomb opening are a starting point for understanding each tomb, how it was laid out and the placement of skeletons and grave goods. The discussions, therefore, are based in part on information taken from the notebooks. There are problems, however, for several reasons. All excavation notebooks are required to be lodged in the Ephoria/Directorate of Antiquities under whose jurisdiction the excavation took place, and they are not allowed to be removed. Between 1969 and 2023 the Ephoria offices were moved from Chania to Rethymnon, with the result that some notebooks or sections thereof were lost. Additionally, the notes were taken by several different assistant archaeologists and this has resulted in a great deal of variation in format and in detail. Drawings were made and they too vary in type and execution and many have not survived. Yannis Tzedakis, assisted by Irini Gavrilaki, located and translated all the notes and drawings that have survived. Where the drawing of a tomb discussed here and in Chapter 9 survives it is reproduced.

Stable isotope analysis

Diet

As reported in Chapter 4, Michael Richards analysed 84 bone samples and nine enamel samples from 18 tombs (Tables 4.1 and 4.2). He concluded that, although the tombs and grave goods indicate a stratified society, the population shared a similar diet, high in protein, meat or dairy, with no quantifiable marine protein and no real difference between the diets of male and female. This was the same conclusion reached by Richards in the initial project (1997–2003) (Richards and Hedges 2008).

The Huddersfield-Oxford group (Chapter 5) reported that 41 of 55 samples of human skeletal material (teeth and petrous bones) produced enough well preserved collagen to give results. Stable isotope results indicated a a similarly broadly homogeneous terrestrial-based diet with no significant amount of marine intake and no real difference between the diets of male and female. Interestingly, in contrast, the study of faunal remains from the tombs and the nearby town indicated that herbivores raised locally were grazed in a different, potentially more wooded or enclosed environment than the fauna represented in the Necropolis, suggesting that the people buried at the Necropolis were not sourcing their food from the town site (see Chapters 5 and 8).

Outliers

(incorporating skeletal summaries by Darlene Weston)
In Chapters 4 and 5 outliers are defined as persons who were born and spent their early years and part of their adulthoods outside Crete. The three identified individuals were men who settled in the 'city' of Armenoi and were then integrated into society. It is presumed this is because they married into a family and were subsequently buried in the family tomb. This is exactly what happens in Greek Orthodoxy today. Men who marry into a family are buried with the wife's family. It is extraordinary that this appears to have been the practice in Minoan society, as evidenced at Armenoi. The three identified outliers are 55Γ and 89B, identified through stable isotope analysis (Chapter 4, Figs 4.1 and 4.2; Table 4.3) and 149 E, identified through ancient DNA analysis (Chapter 5). Outliers were not unusual in the Necropolis, according to Richards (Chapter 4). He found evidence possibly, but not conclusively, of at least four others (69A, 76Δ, 95λ and 146A). The osteological reports below on skeletons 146A, 55Γ and 89B (discussed in that order) have been prepared with data recorded according to the standards outlined by Buikstra and Ubelaker (1994) and Weston (2019).

Tomb 146: Skeleton 146A

Biological sex: possible male
Age at death: 46+ years
Date: LM IIIA–LM IIIB

Skeleton 146A is from a tomb that is especially important because it contained the stirrup jar with the Linear B inscription (see Chapter 9, Fig 9.39). Therefore, the possibility of an outlier in the tomb is significant. The skeleton was of a possible male, aged over 46 years. It was only 25–50% complete and the preservation was poor, which limited the ability to determine stature (for details see Chapter 3). Elements present included fragments of the mandible, arm and leg bones, vertebrae, pelvis, and hands and feet. Most of the maxillary dentition was destroyed, but that which remained included heavily worn left 1st and 2nd molars, with complete destruction of the crowns. Unusually, there was very little wear on the mandibular canines, asymmetric wear on the left mandibular 2nd incisor, congenital absence of mandibular central incisors, and a very small left mandibular 1st premolar. There were no skeletal pathologies present. For additional information on Tomb 146 and its contents see Chapter 9.

Tomb 55: Skeleton 55Γ

Biological sex: male
Age at death: 46+
Date: Late Minoan IIIB:1–IIIB:2 (ca. 1340–1190 BC)

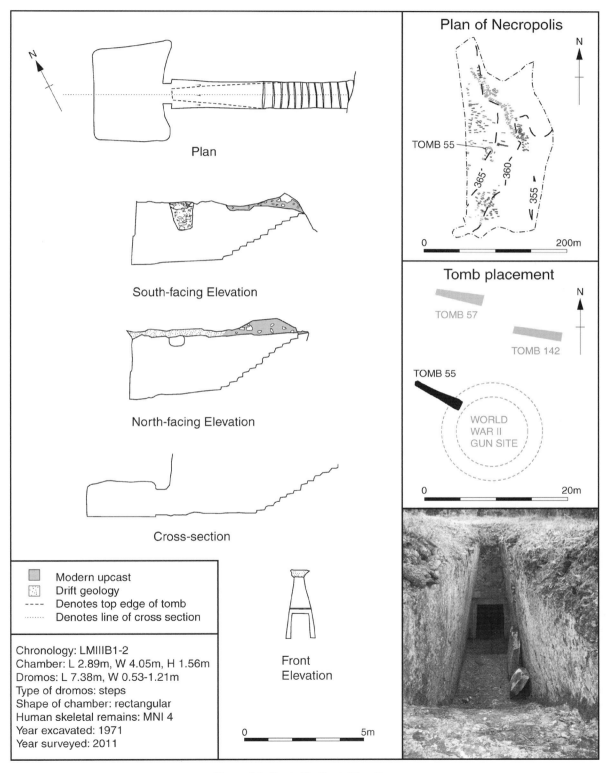

Figure 6.1. Tomb 55: Tomb Plan Page

Figure 6.2. Larnax A (R.M. 1703) Tomb 55, with gabled lid, decorated with stylised octopi on long sides and horns of consecration on a short side. LM IIIB.

R.M.M. 347

Figure 6.3. Bronze cleaver (R.M.M 347). Tomb 55. LM IIIB:1.

R.M.M. 385

Figure 6.4. Bronze dagger (R.M.M. 385). Tomb 55. LM IIIB:2.

Tomb 55 (Fig. 6.1) is one of the richest tombs with two larnakes. It had a lifespan of ca. 100 years. In World War II the Germans put an anti-aircraft gun emplacement over this well-made, rich tomb with its beautifully carved entrance. Thankfully they never discovered what lay beneath it. In the space between the larnakes two skeletons lay on the floor of the chamber. 55Γ was a male aged at least 46 years or older at the time of death. The stature of the individual is estimated to be 163.16±4.05 cm, and the preservation of the skeleton was good. Skeletal completeness was 50–75%, and bones

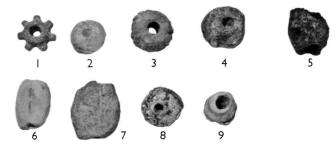

Figure 6.5. Rich selection of small finds (beads and jewellery) in Tomb 55. LM IIIB: 1. faience with relief bands; 2. rock crystal; 3–4. faience with incised lines; 5. Barrel shaped carnelian; 6–7. biconical faience; 8. faience with incised lines; 9. glass paste.

of the calvarium, arms, ribs, ossa coxae and femora were present. The dentition exhibited ante-mortem tooth loss in the maxillae, particularly affecting the molars and premolars. This suggests that the individual had poor dental health, as ante-mortem tooth loss is usually a result of untreated caries, abscesses/granulomata, or periodontal disease. As the mandible was not present, it was not possible to find out if similar dental pathology was observable in the lower jaw. The only skeletal pathology observed was a rib fragment with an unusual, curved head articulation, suggestive of a long-standing rib fracture.

Larnax A (R.M. 1703) dates to Late Minoan IIIB (Fig. 6.2) (Tzedakis and Kolivaki 2018, fig. 1.24) and Larnax B (R.M. 1846, not illustrated) is dated to Late Minoan IIIB:2. The long sides are decorated with wavy lines in imitation of octopus tentacles. The tomb also contained fine pottery, bronzes, and many small finds (see below and Appendix). Close to the skull of 55Γ were a bronze cleaver (R.M.M. 347) (Fig. 6.3) and dagger (R.M.M 385) (Fig. 6.4), both of Mycenaean type. Nails (presumably from a wooden bier) also lay close by. The burial of a child found at the feet of the Skeleton 55 Γ was possibly a child of, or related to, him.

There was a rich selection of small finds (beads and jewellery) found in the tomb; the most interesting beads are illustrated (Fig. 6.5). Additional small finds of note include: two seals, a wire bracelet with coiled wire over the join; a silver ring; two nails; and fragments of lead. 55Γ could have been a trader or settler or, considering the artefacts that had been placed near the top of the skull, was more likely to have been a warrior. Could he have been a warrior integrated into Minoan society? Could he have been a Mycenaean? There were Mycenaean products in the tomb. This could have been the result of trade, but the finds near the skeleton indicate that this man is more likely to have been a Mycenaean.

For the catalogue entries for Tomb 55 see Appendix.

Tomb 89: Skeleton 89B
Biological sex: male
Date: Late Minoan IIIA:1–Late Minoan IIIA:2 (ca. 1390–1340 BC)
Age at death: 26–35

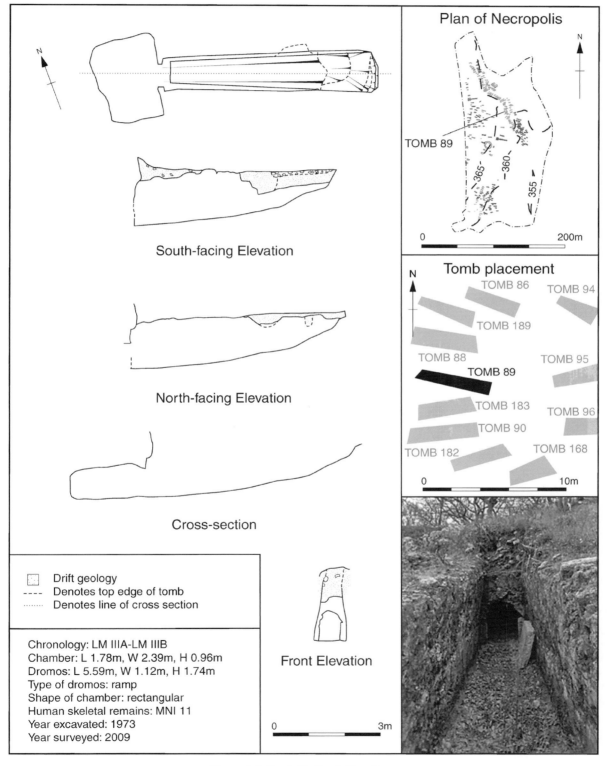

South-facing Elevation

North-facing Elevation

Cross-section

Plan of Necropolis

TOMB 89

0   200m

Tomb placement

TOMB 86   TOMB 94
TOMB 189
TOMB 88   TOMB 95
TOMB 89
TOMB 183   TOMB 96
TOMB 90
TOMB 182   TOMB 168

0   10m

::: Drift geology
---- Denotes top edge of tomb
...... Denotes line of cross section

Chronology: LM IIIA-LM IIIB
Chamber: L 1.78m, W 2.39m, H 0.96m
Dromos: L 5.59m, W 1.12m, H 1.74m
Type of dromos: ramp
Shape of chamber: rectangular
Human skeletal remains: MNI 11
Year excavated: 1973
Year surveyed: 2009

Front Elevation

0   3m

Figure 6.6. Tomb 89: Tomb Plan Page.

Figure 6.7. Beak-spouted, decorated jug (R.M. 2144). Tomb 89. LMIII A:1/IIIA:2.

Figure 6.8 Beak-spouted decorated jug (R.M. 2148). Tomb 89. LM IIIA:2.

Tomb 89 (Fig. 6.6) is one of the earliest tombs containing the remains of 13 skeletons. No specific information survives on their placement in the tomb. Skeleton 89B belongs to a male, aged 26–35 years. With a height of 176.65±3.27 cm, this individual was the tallest interred in the tomb. The preservation of this skeleton was good, with 75–100% completeness, missing primarily the hands and feet. The dentition of the individual showed ante-mortem loss of the left maxillary molars and most of the mandibular molars, with the alveolar bone in a state of remodeling and porosity of the palate. The tooth loss was likely a consequence of severe caries or periodontal disease. This individual had an atypical vertebral column, with 14 thoracic vertebrae (two extra) and four lumbar vertebrae (one less than normal) present. Evidence for degeneration of the spine, including a Schmorl's node was present which, due to the individual's relatively young age, is likely indicative of the performance of strenuous physical activity.

In spite of the large number of skeletons compared with others in the Necropolis Tomb 89 was not rich in artefacts. Of the nine vases found five are undecorated and from unknown workshops; one from the Armenoi workshop had traces of paint. There are three decorated vessels, one painted with concentric circles and two with an iris motif. These three were products of the Knossos workshop.

According to Jan Driessen (1990), the Mycenaean invasion of Crete took place in the Late Minoan II period, ca. 1425–1390 BC and the fall of Knossos dates to Late Minoan IIIA:2, ca. 1370–1340 BC. Tomb 89 was built before the fall of Knossos. A beak-spouted jug (R.M. 2144) (Fig. 6.7) and a miniature jug (R.M. 2148) (Fig. 6.8) are examples of the vases found in the Necropolis which were made at the palace workshop of Knossos in Late Minoan IIIA:1/IIIA:2.

The presence of these vases indicates that, at this time, the 'city' of Armenoi enjoyed good relations with Knossos. The impact of its fall on the inhabitants of Armenoi can only be conjectured. It had to have one, but there is no evidence of any hiatus in the building or the use of tombs, which there would have been if the impact had been great. This raises many questions, including the exact nature of the relationship Armenoi had with Knossos before, at the time of, and after the fall. The only concrete statement that can be made is that it appears the pottery workshops at Knossos were in operation throughout the period of its demise and that there was continuity of culture at the site of Armenoi.

For the catalogue entries for Tomb 89 see the Appendix.

Evidence for migration

A study published in 2022 (Richards *et al.* 2022) gives fascinating results on the varied lives of two people (76Δ male and 95Λ female) who were buried in the Necropolis in two other tombs, which are not a result of the analyses carried out under the auspices of this project. Sulphur values indicated their adult lives were spent in great part outside Crete. Skeletons 55Γ and 89B have both strontium and sulphur isotope values that are higher than the faunal averages for Crete and most of the other individuals from Armenoi (Table 4.3) indicating that much of their adult lives were also spent away from Crete.

The ancient DNA work undertaken by Foody *et al.* (Chapter 5) further identified an outlier whose ancestry may have lain in Western Europe (see below). These few cases indicating possible migration is significant given the small proportion of the overall number of burials sampled from the Necropolis and limited number of tombs from which they came and indicates that the burial of individuals of foreign descent or domicile in family tombs was accepted practice by the Minoans and a recognition of the important roles they must have played in Minoan society. The population in the 'city' of Armenoi was clearly a cosmopolitan one.

Ancient DNA

Outlier

The DNA results indicate that Skeleton E from Tomb 149 is clearly an outlier, who plots closer to Bronze Age individuals from Western Europe than from Crete itself. In contrast, the PCA plot of the other 22 individuals from the Armenoi that were analysed for DNA indicate an homogeneous population. They overlap both Mycenaeans from mainland Greece and, to a slightly lesser extent, the earlier Minoans (Early Minoan I–Middle Minoan IIB) from other parts of Crete, all of which derive much of their ancestry from the Anatolian Neolithic which persisted through into the Bronze Age (Chapter 5).

The in-depth study of the pottery from the Necropolis presently being undertaken by Tzedakis and Kolavaki (in prep.), and Martlew's study of Minoan and Mycenaean pottery (in prep.), support this view. The pottery is mainly Minoan. There are a very few Mycenaean imports and a few cases made in the Knossos workshop imitate Mycenaean prototypes. The majority of the bronzes imitate Mycenaean prototypes. The chamber tomb was a Mycenaean innovation which was copied in Crete. The larnakes are Minoan.

Tomb 149: Skeleton 149 E

Biological sex: male

Date: Late Minoan IIIA:2–Late Minoan IIIB:1 (ca. 1370–1250 BC)

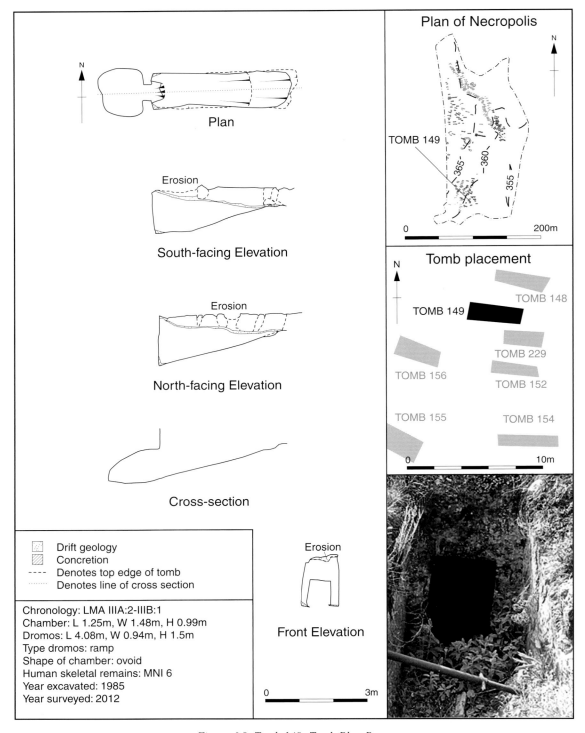

Figure 6.9. Tomb 149: Tomb Plan Page.

This tomb (Fig. 6.9) contained six skeletons. Unfortunately, a sample taken from 149E for radiocarbon dating failed to produce a date due to low collagen. What we do know, from a study of the pottery, is that the tomb had a long lifespan, being built in the second period after the foundation of the Necropolis, in ca. 1370 BC. It was in use until ca. 1250 BC. A radiocarbon sample on one of the other skeletons (ΣT) produced a date of 1432–1282 cal BC (95.4%; Table 5.3). The finds included fine pottery and a bronze spearhead.

Six things stand out about the tomb which tell us about the lifestyle of its citizens:

1. The vases were in good to pristine condition. 'New' vases in the tomb indicate how much members of families were revered and how important it was to have the finest in the tombs for funerary ritual. This is obvious because of the very existence of the Necropolis but it shows that caring continued down to details, in spite of the fact that old burials were pushed into heaps in the corners of chambers as part of accepted ritual.

2. The presence of two vases from Knossos, a miniature stirrup jar (R.M. 3407) dated to Late Minoan IIIA:2 (Fig. 6.10) and an alabastron (R.M. 3396) dated to Late Minoan IIIB:1 (Fig. 6.11), and those found in other tombs, stand witness to the fact that the 'city' of Armenoi enjoyed a healthy trade with the pottery workshop at the Palace of Knossos at a time just before its final destruction in Late Minoan IIIA:2 in ca. 1370–1340 BC as well as afterwards. Once again it raises a host of intriguing questions about the exact relationship between the city of Armenoi and Knossos at this pivotal moment in time.

3. Finds indicate the amount of trade that existed between the 'city' of Armenoi and Kydonia. Examples are a stirrup jar (R.M. 3390) dated to Late Minoan IIIA:2 (Fig. 6.12)

and a squat-conical stirrup jar (R.M. 3388), dated to Late Minoan IIIB, which is in such perfect condition it looks newly made (Fig. 6.13), both from the Kydonia workshop.

4. Beautiful vases were made in the Armenoi workshops (Tzedakis has identified two) and their presence in the tomb indicates that local production was much valued

Figure 6.11. Alabastron from the Knossos workshop (R.M. 3396). Tomb 149. LM IIIB:1.

Figure 6.10. Miniature stirrup jar from the Knossos workshop (R.M. 3407). Tomb 149. LM IIIA:2.

Figure 6.12. Stirrup jar decorated with stylised octopus tentacles. Kydonia workshop (R.M. 3390). Tomb 149. LM IIIA:2.

by its citizens. the identified examples are a jug (R.M. 3416) dated to Late Minoan III B middle (Fig. 6.14) and an alabastron (R.M. 3422) dated to Late Minoan III A:2 (Fig. 6.15).

5. An outstanding find is an ivory comb (R.M.O. 1085) dated to Late Minoan IIIA. It was found in two fragments, the largest 3.4 cm in length (Fig 6.16).

6. Perhaps most surprising of all, the family in this tomb welcomed a foreigner into its bosom and allowed him to be buried with its loved ones.

For catalogue entries for Tomb 149, see Appendix.

Figure 6.13. Squat-conical stirrup jar, so perfect it looked newly made. Kydonia workshop (R.M. 3388). Tomb 149. LM IIIB:1.

Figure 6.14. Jug decorated with concentric bands. Armenoi workshop (R.M. 3416). Tomb 149. LM III B.

Figure 6.15. Alabastron decorated with papyrus. Armenoi workshop (R.M.3422). Tomb 149. LM III A:2.

Figure 6.16. Ivory comb (R.M.O. 1085). Tomb 149. LM IIIA.

Sex determination

Twenty-three individuals were analysed using in-depth NGS (Chapter 5, Table 5.2). These were sexed genetically indicating 16 females and seven males were identified from ten tombs. In addition, osteological analysis (Chapter 3) identified four males from Tomb 159.

Kinship analysis

The ability to establish kinship in the chamber tombs was considered the most important outcome of the ancient DNA programme. It depended wholly on whether sufficient DNA had been retained in any of the samples taken from skeletal material found in a Bronze Age Necropolis that dated ca. 1390–1190 BC. This was very problematical. What gave us hope was that DNA had been retained in a sample taken earlier from a female femur from Tomb 160 (Lazaridis and Mittnik *et al.* 2017).

Finally it depended on the submission of more than one sample from a tomb which had equally positive results. This made the situation even more problematical. The total number of samples and those sent for full sequencing = 171 samples from 118 individuals from 48 tombs; 55 samples were extracted, 41 from 16 tombs had preserved collagen which were then screened for ancient DNA. From these, 23 samples from ten tombs were submitted to full DNA sequencing (see Chapter 5 for details):

Tomb 67, n = 1, Skeleton E
Tomb 69, n = 1, Skeleton Γ
Tomb 78, n = 1, Skeleton Z se
Tomb 149, n = 3, Skeletons Δ, E and ΣT
Tomb 167, n = 2, Skeletons E and Z
Tomb 198, n = 3, Skeletons B, Γ and Δ
Tomb 203, n = 8, Skeletons A, E, Z, H, Θ, I, K and ΣT
Tomb 204, n = 1, Skeleton Z
Tomb 208, n = 1, Skeleton E
Tomb 210, n = 2, Skeletons E and 4

Out of only four tombs from which more than one sample could be submitted, three were successful: Tombs 198, 203, 210 demonstrated kinship (see Chapter 5, Table 5.2). Given the small size of the sample from the amount of skeletal remains that survived in the Necropolis, these results are of even greater significance.

Tomb 198: Skeletons 198B, 198Γ and 198Δ
Biological sex: female and ?female
Date: Late Minoan IIIA:2–Late Minoan IIIB:2 (ca. 1370–1190 BC)

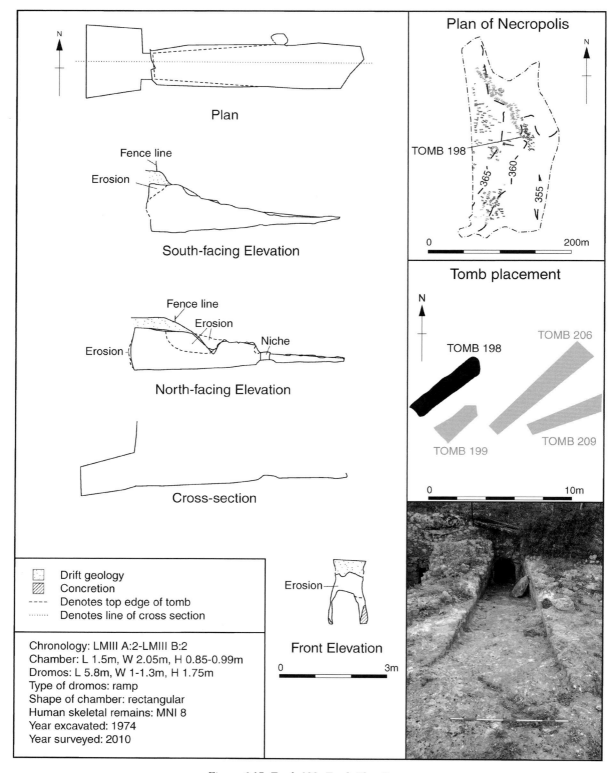

Plan

Fence line
Erosion
South-facing Elevation

Plan of Necropolis
TOMB 198
0   200m

Fence line
Erosion
Erosion
Niche
North-facing Elevation

Tomb placement
TOMB 198
TOMB 206
TOMB 199
TOMB 209
0   10m

Cross-section

Drift geology
Concretion
- - - Denotes top edge of tomb
· · · · · Denotes line of cross section

Chronology: LMIII A:2-LMIII B:2
Chamber: L 1.5m, W 2.05m, H 0.85-0.99m
Dromos: L 5.8m, W 1-1.3m, H 1.75m
Type of dromos: ramp
Shape of chamber: rectangular
Human skeletal remains: MNI 8
Year excavated: 1974
Year surveyed: 2010

Erosion
Front Elevation
0   3m

Figure 6.17. Tomb 198: Tomb Plan Page.

Tomb 198 (Fig. 6.17) is one of the richest tombs with continuous occupation over a long period from very near the beginning, 1370–1340 BC, until the end of use of the Necropolis. The original drawing, made when the tomb was opened, has survived (Fig. 6.18). The tomb presents a good example of burial practice at the Necropolis. Eight skeletons were found. Older burials were pushed aside when a new one took place. This is clear in the drawing. The latest secondary burial, Skeleton 1, is drawn on the left side near the entrance. The skull (A) was found at the angle of the entrance. Two decorated cups (R.M. 6588 and R.M. 6589, a miniature) were found with Skeleton 1. Both date to Late Minoan IIIB and R.M. 6589 specifically to Late Minoan III B:2, the end date of use of the tomb and the Necropolis. All three skeletons sampled were earlier burials that had been pushed into the north-west, upper right corner (facing).

The results reveal that the family buried in the tomb was a wealthy and presumably influential one, and certainly long lived. Of the three females, 198B and 198Γ showed a first-degree relationship and were either mother

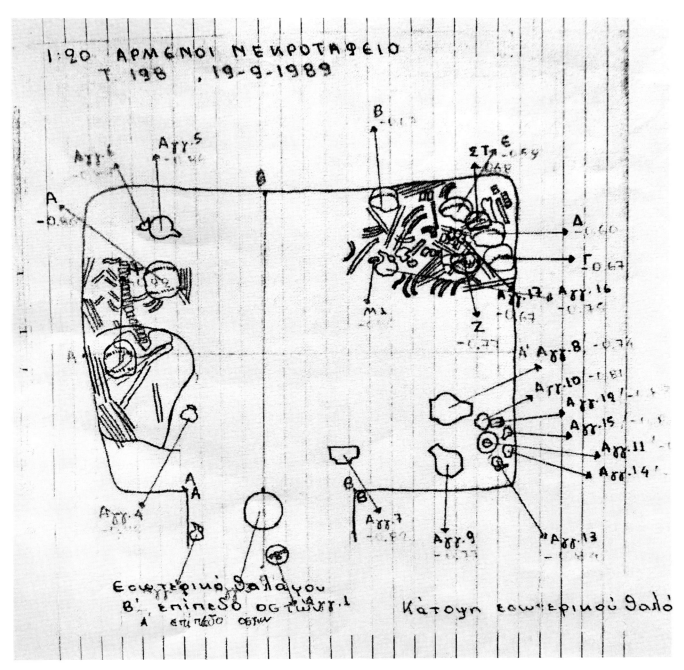

Figure 6.18. Drawing of Tomb 198 from the Excavation Notebook.

Figure 6.19. Double vase. Two miniature jugs joined, decorated (curved stripes FM.67). One handle has an animal protome. Armenoi workshop (R.M. 6600). Tomb 198. LM IIIA:2.

Figure 6.20. Jug. Beak-spouted, decorated (scale pattern FM.70). Armenoi workshop, imitation of Knossos prototype (R.M. 6588). Tomb 198. LM IIIA:2.

Figure 6.21. Decorated mug with handle (R.M. 6589). Tomb 198. LMIIIB:2.

and daughter, or sisters; 198Δ also female(?), found next to 198Γ, was not related to either of the others.

Two vases found in the tomb were found near Skeleton 198 Γ. They date to the same period, Late Minoan IIIA:2, during which time the tomb was built (1370–1340 BC). It was hoped that pottery found near the skeletons would reveal something concrete about their relationship. This is a period of 30 years, so these two females could either be sisters or mother and daughter, as indicated by DNA analysis, but it is more likely that they were the latter.

Found in the same area which held the remains of 198B and 198Γ was a double vase (R.M. 6600) from the Armenoi workshop – two miniature jugs joined, decorated, and one handle has an animal protome – dated to Late Minoan IIIA:2 (Fig. 6.19). This is an important find. As a double vase with a handle which features an animal protome, it is like that cited by Godart and Tzedakis (1992, pl. LII; see Fig. 10.9), which Tzedakis found in a chamber tomb at Kalami (on the main road between Rethymnon and Chania), so it could provide evidence for commerce between the 'city' of Armenoi, Pylos and Mycenaean sites on the Mainland (see Chapters 9 and 10).

Vases found in the tomb included a beak-spouted jug (R.M. 6588) (Fig. 6.20), and a mug with a handle (R.M. 6589) (Fig. 6.21), both dated to Late Minoan IIIB:2. Taking into consideration the beautiful vases lying with them and the abundance of jewellery, including hair rings, 198B and 198Γ were two very high status females.

For the catalogue entries for Tomb 198 see Appendix.

Tomb 210: Skeletons 210E and 210-4
Biological sex: female
Date: Late Minoan III A–Late Minoan III A:2/Late Minoan IIIB:1 (ca. 1390–1250 BC)

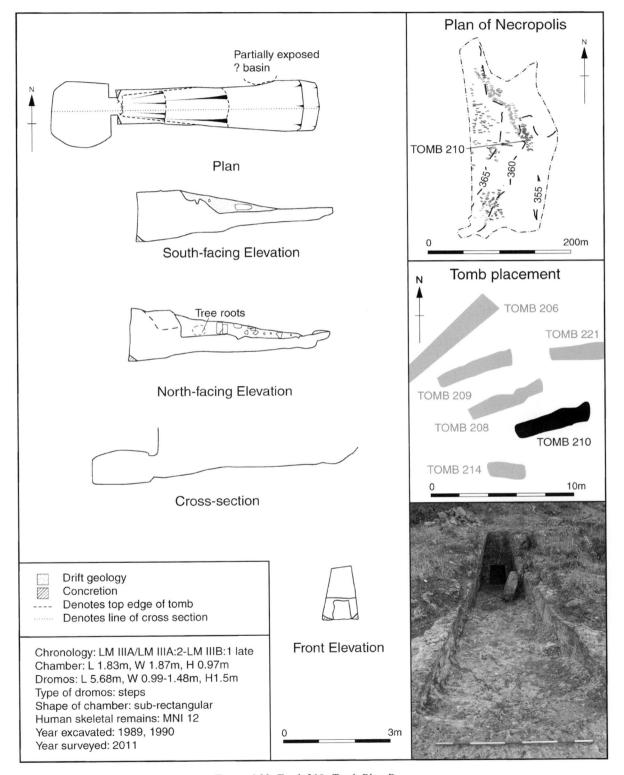

Figure 6.22. Tomb 210: Tomb Plan Page.

The remains of 12 skeletons were found in Tomb 210 (Fig. 6.22). Eleven were on the right facing/north side of the chamber. One burial was found on its own, *in situ* and would be the latest. Skeleton 210-4 was radiocarbon dated to 1421–1263 (95.4%). The vases found with 11 skeletons are dated to Late Minoan IIIA. The pottery near the *in situ* skeleton is Late Minoan IIIB:1. This indicates that Tomb 210 was in use for less time than Tombs 198 and 203.

These remains were found with vases dated to Late Minoan IIIA, ca. 1390–1370 BC. These two individuals had a first-degree relationship. Skeleton 210-4 is a child and she could be the sister, but more likely the daughter, of Skeleton 210E. The fine vases found in the tomb, dated to Late Minoan IIIA included: R.M. 7673, a miniature alabastron (Fig. 6.23); R.M. 7664, a bridge-spouted jug (Fig. 6.24); R.M. 7672, a beak-spouted jug (Fig. 6.25); R.M. 7663, a piriform jar (Fig. 6.26); R.M. 7668, a beak-spouted jug (Fig. 6.27).

For the catalogue entries for Tomb 210 see Appendix.

R.M. 7672

Figure 6.25. Beak-spouted, jug decorated with quirk pattern (R.M. 7672). Tomb 210. LM IIIA:2.

R.M. 7673

Figure 6.23. Miniature 3-handled globular alabastron (R.M. 7673). Tomb 210. LM IIIA:1– LM IIIA-2.

R.M. 7663

Figure 6.26. Piriform jar decorated with elaborate triangles (R.M. 7663). Tomb 210. LM IIIA:2.

R.M. 7664

Figure 6.24. Bridge-spouted net pattern jug (R.M. 7664). Tomb 210. LM IIIA:2/IIIB:1.

R.M. 7668

Figure 6.27. Beak-spouted jug decorated with quirk pattern (R.M. 7668). Tomb 210. LM IIIB.

Tomb 203: Skeletons I and A

Biological sex: Skeleton I: male; Skeleton A: female
Date: Late Minoan IIIA:2–Late Minoan IIIB:2 (ca. 1370–1190 BC)

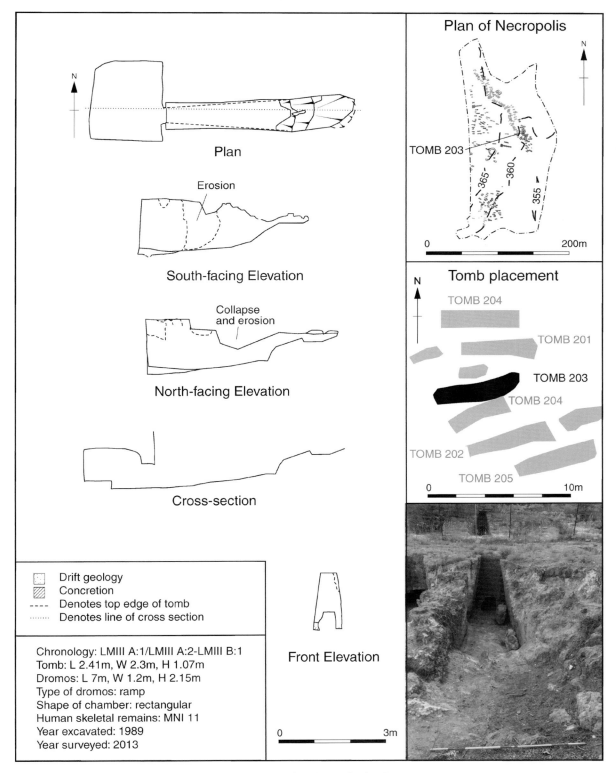

Figure 6.28. Tomb 203: Tomb Plan Page.

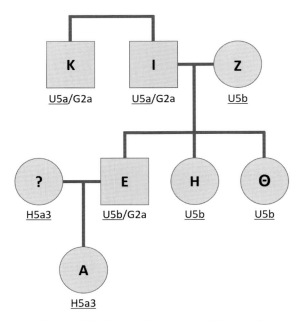

Figure 6.29. Genetic diagram (see Chapter 5).

Tomb 203 (Fig. 6.28) is one of the richest tombs with continuous use over a long period, nearly the full life of the Necropolis. Tomb 203 is the most important of all the tombs from which samples were sent for full sequencing and contained a multi-generational family. The remains of 11 people were found, eight of which were sent for analysis. Seven samples produced results that indicated they had a first- or second-degree relationship. Only female ΣT was not related to any of the others. Skull A was shown to be *third*-generation and, as indicated on the genetic diagram (Fig. 6.29), is indicated to be the paternal granddaughter of Skeleton I (Chapter 5).

This, however, raised a problem. The contents of the chamber were drawn according to where they were found when the tomb was first excavated (Fig. 6.30). Skeleton I is drawn to the left of the entrance to the tomb. The tradition followed was that the latest burial was placed at the front of the tomb. Skull A is clearly marked. The original notes say that Skull A belongs to Skeleton I. In the context of the tomb plan this interpretation makes sense but the ancient DNA indicates otherwise. The DNA result from Skeleton I

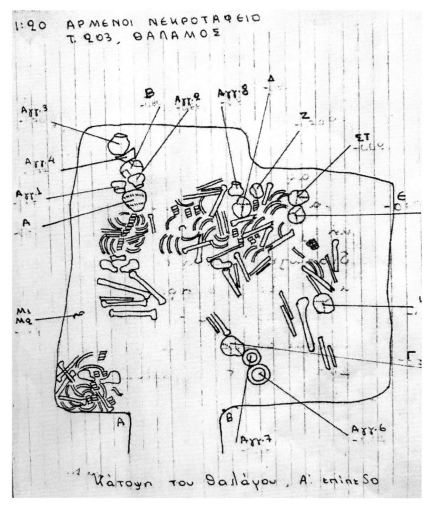

Figure 6.30. Drawing of Tomb 203 from the Excavation Notebook.

comes from the right petrous. This individual is a *male*, with mtDNA haplotype U5a and Y-chromosome haplotype G2a. From Skull A, the left petrous was sampled (Figs 6.31 and 6.32). This individual is a *female* with mtDNA haplotype H5a3. Skull A is, therefore, definitely *not* the same individual as the petrous labelled as being from Skeleton I. The sampling protocol was to take samples where the researchers were sure (as could be) that the samples were from different skeletons. Figures 6.31 and 6.32 are of the two samples. A rechecking of the relevant notebook indicated that the roof of tomb 203 had fallen in which may have disturbed the contents and would account for the relative positions of the two skull fragments. Jar R.M. 7518 was found under the thorax of Skeleton I, together with many bronze fragments, earrings and rings and a beak-spouted jug decorated with concentric semicircles, from the Armenoi workshop (R.M. 7520) that dates to Late Minoan IIIB:1 (Fig. 6.33).

An alabastron (R.M. 7517 was found beside Skull A. It is an imitation of a Mycenaean prototype from the Knossos workshop. (Fig. 6.34) Also found was a piriform jar (R.M.7518), also an imitation of a Mycenaean prototype made in the Knossos workshop and dated to Late Minoan IIIA:2 (Fig. 6.36). These two vases illustrate once more the influence Mycenaean vessels had on the potters at the palace site of Knossos. The finding of an alabastron is worth noting, as it is an imitation of a Knossos prototype made in an Armenoi workshop, dated to Late Minoan IIIA:2. The influence Mycenae exerted upon Knossos, and Knossos exerted upon Armenoi, is intriguing. It begs a lot of questions and is worthy of a study on its own. R.M. 7519 is a cup from an Armenoi workshop (Fig. 6.35), dated to Late Minoan IIIA:2. R.M. 7521 (Fig. 6.37)., is an alabastron from the Kydonia workshop, dated to Late Minoan IIIB:1.

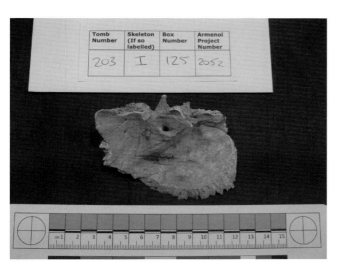

Figure 6.31. Sample from Skeleton I. Tomb 203.

Figure 6.32. Sample from Skull A. Tomb 203.

Figure 6.33. Beak-spouted jug decorated with concentric semicircles. Armenoi workshop (R.M. 7520). Tomb 203. LM IIIB:1.

Figure 6.34. Globular alabastron decorated with papyrus. Knossos workshop, imitation of Mycenaean prototype (R.M. 7517). Tomb 203, found beside Skull A. LM IIIA:1–LM IIIA:2.

Figure 6.35. Cup decorated with concentric semicircles. Armenoi workshop. Tomb 203 (R.M. 7519). LM IIIA:2.

Figure 6.36. Piriform jar. Imitation of a Knossian prototype (R.M.7518). Tomb 203. LM IIIA:2.

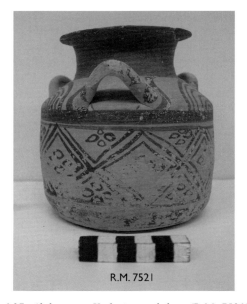

Figure 6.37. Alabastron. Kydonia workshop (R.M. 7521). Tomb 203. LM IIIB:1.

This vase provides additional evidence of trade between Armenoi and Kydonia. However, because of the collapse of the ceiling of Tomb 203, the context of the vases cannot be securely identified as being of the same dates as Skeleton I or Skull A.

For the catalogue entries for Tomb 203 see Appendix.

Kinship revealed by osteological analysis

Weston's analysis of the skeletal remains from Tomb 159 revealed a familial relationship between two of the males buried within the 'royal' tomb, another important and exciting discovery (see Chapter 3).

Tomb 159: Skeletons 159Γ and 159Δ

Biological sex: male
Age at death: 159Γ: 26–35; 159Δ: 36–45
Date: Late Minoan IIIA;1–Late Minoan IIIB1 (ca. 1390–1250 BC)

The minimum number of individuals in this tomb is five comprising two male adults, two possibly male adults and a foetus. Of these, two had the same unusual variety of pronounced overbite where the lower teeth are essentially covered, abraded, and worn by the upper teeth (see Chapter 3). Skeleton 159Γ was aged 26–35 years and 159Δ was 36–45. Were they brothers or father and son? According to Weston, being father and son is more likely (Chapter 3). What osteological analysis revealed in general about the inhabitants in the royal tomb, is important. To be able to establish kinship between two of the inhabitants, even moreso. Because of what it is (the royal tomb) and who it is (certainly the ruling family, one ruler or two) it is one of the most important and exciting discoveries presented in this volume.

Radiocarbon dating

Only a limited suite of radiocarbon dates could be obtained from individuals sampled for DNA and/or staple isotope analysis and the results fall at an awkward part of the calibration curve where date ranges tend to be broadened. Two samples could not be dated due to low collagen levels. Consequently the results are of limited value in nuancing the use of the individual tombs. The combined calibrated date range of 1409–1266 cal BC at 95.4% confidence from skeletons in Tomb 149 is in broad agreement with the dating based on ceramic typology, that the life of the Necropolis of Armenoi was between 1390 and 1190 BC.

Conclusions

Trade and influence

Pottery is the great indicator of trade and influence. Trade between sites matters, influence between sites, especially between Mycenaean sites and Crete, matters. Significant in the context of Armenoi is the openness of the potters of the 'city' of Armenoi to new ideas. If they liked something, they had no qualms about adding it to their repertoire. This is illustrated by alabastron R.M. 7518 from Tomb 203 (Fig. 6.36), which is an imitation of a Knossos Late Minoan IIIA:2 prototype. This stands as an insight into the minds of Minoans at the time. They were not hidebound. They were receptive to the art of others.

Minoan Society: what has been revealed?
A 'king'
The inhabitants of the chamber tombs lived under a king or ruler who was buried in great style, as revealed by the contents of Tomb 159 (see Chapters 3, 9 and 10)

Tombs: elaborate and simple (ritual)
There were rich tombs as well as some less rich ones. They had large enough chambers and dromoi, both replete with vases, that give credence to the idea that both were used to carry out impressive funerary rituals. The remains of a ritual feast were found in the dromos of Tomb 159, the royal tomb (see Chapter 9).

There were also simple tombs with chambers so small that Professor Robert Hedges could hardly squeeze himself inside when attempting to acquire DNA samples. This did not preclude ceremonies taking place in the dromos of a simple tomb and there were examples that contained pottery suitable for such an occasion. It is believed that society was stratified, but rich and the less well-off alike had the right to be, and were, buried all together on the same hillside, and the majority of tombs and/or dromoi contained the types of vessels which suggest ritual use.

Equality in society
Foody *et al.* in their DNA analysis reported in Chapter 5 concluded, albeit on the basis of a comparatively small sample, that it was not possible to identify evidence of a patriarchal society from the genetic analyses, but that the finding of more females than males (16 and seven, respectively) may indicate that Late Minoan III Crete was not a patriarchal society.

There is one obvious exception of what appears to be relative equality between males and females in the chamber tombs. The royal tomb, 159, had 4 males and one foetus (not possible to sex). There is no easy explanation for this. One is that all the skeletal material did not survive, due to burial in limestone, climate, and in the case of Tomb 159, tomb robbery at the end of the Minoan period which would have exacerbated the exposure of the skeletal remains to the elements. Female bones were smaller and more fragile. Due to all these factors, there were no complete skeletons found in the Necropolis. Although rich and poor tombs do indicate stratification in society, it is not decreed, it probably depended on family circumstances, and even this could be part of the reason for the sex of the inhabitants

found in Tomb 159. The fact is that the rich and the less well off were both allowed to have chamber tombs built in the same Necropolis, and this is the accepted norm from the time the Necropolis was founded. This indicates a type of equality in society. With reference to diet, there was no discernible difference male and female or between rich and poor. Taking all the above into account, it can be concluded that there was a great deal of equality in Minoan society in Late Minoan III. This is a broad generalisation, but it cannot be that the 'city' that built the Necropolis stood alone in its beliefs and mores.

Diet and cooking pots
There was no discernible difference in diet. It was terrestrial-based, high in protein (meat or dairy) and did not contain any traceable amount of seafood. This is the custom to this day. Cretans who live in the mountains either do not eat fish at all or only rarely. What is so strange is that most Minoans lived along the shore or within reasonable distance from it, many were seafarers, and some of the finest pottery the Minoans ever produced is exquisitely decorated with marine life. Octopi are one of the main motifs the Minoans painted on the larnakes in which they found their final resting places, their tentacles' undulating along the long sides (see back cover). Why? What were the beliefs behind this remarkable situation? What was the reasoning behind it? We have no answers to these questions.

Although living in a period after the Mycenaeans invaded Crete, the scientific analyses indicate that the inhabitants of the tombs were predominantly Minoan. The excavator, Yannis Tzedakis, now has proof for what he had suspected all along from his study of the vases found at the Late Minoan III Necropolis, and on good grounds. The same is true for coarseware pottery. Tripod cooking pots found in western Crete a few of which are Mycenaean in shape, are large like traditional Minoan cooking pots. Mycenaean ones on the Mainland, with one or two exceptions found in ritual locations, are tiny in comparison. Mycenaean influence is clear, but the potters had to be Minoans whose clients were Minoan cooks who were preparing food in the traditional way.

The outliers and travellers
Skeletons 55Γ, 89B, and 149E: Imagine three men who were born and lived elsewhere, settled in the wealthy 'city' which was built on the main north–south trade route in West Crete. They came for whatever reason, perhaps as traders, seafarers or warriors, and each fell in love, took a Minoan bride and was buried with her in a splendid chamber tomb. For the very first time in the history of the Minoan civilisation, through stable isotope analysis and ancient DNA, and the information contained in the original excavation notebooks, we have been able to identify three non-Minoans who 'married' into wealthy families, were integrated into Minoan society, and were buried with their adopted families.

Were they Mycenaeans?: while neither isotopes nor DNA can be definitive, based on current evidence it is reasonable to infer that outliers 55Γ and 89B were, or could have been, Mycenaean. Intriguingly 149E could *not* have been Mycenaean as DNA indicated a Western European origin.

Skeleton 76Δ (male) and Skeleton 95Δ (female): Stable isotope analysis results indicated there were some individuals buried in the Necropolis who had lived some of their lives outside Crete (Chapter 4; Richards *et al.* 2022). Taken with the evidence of outliers (those *born* outside Crete), it has to be said that the 'city' had thoroughly integrated foreigners and well-travelled people into their society. The 'city' of Armenoi was indeed a cosmopolitan one.

Families

Families, above all, were what we hoped to find, and we did. Although the sample size was small in comparison to the minimum number of individuals known to be buried in the Necropolis, families were identified by ancient DNA in Tombs 198, 203 and 210, and by osteological analysis in Tomb 159, the royal tomb. The identification of a probable father and son in the royal tomb, a mother and daughter in Tomb 198, three generations of one family in Tomb 203, and the possibility of a mother and baby in Tomb 210 are hugely significant in beginning to unravel the ancestral and familial complexities of Bronze Age society in the eastern Mediterranean.

What we have learned

When applying scientific analyses to archaeological discoveries, ceramic and human skeletal material, all avenues must be explored. Here the combined results of organic residue, stable isotope, osteological, ancient DNA and radiocarbon analyses have provided both detailed and tantalising evidence concerning the population of the 'city' of Armenoi and its Necropolis. Establishing the existence of families buried in the tombs brings the Necropolis to life, makes us view the inhabitants in the round, as real people, in a way that nothing else could do. To accomplish this was a triumph for the Late Minoan III Necropolis of Armenoi project and for science.

The future

It is important to remember that by no means has all the skeletal material found in the chamber tombs of the Necropolis been scrutinised and submitted to scientific analysis. In the case of ancient DNA, 62 samples taken in 2017 remain to be analysed and only a few of the tombs have undergone in-depth osteological analysis. The Late Minoan III Necropolis of Armenoi has proved itself in two science projects. The success is clear. The Necropolis is a perfect subject for additional, advanced, state of the art, scientific research. There is a future. How exciting it is that there is clearly more work to be done, and that should be done.

Unlocking Secrets

Unlocking secrets is what this book is all about. The results presented in Chapters 3, 4, and 5 are astonishing. The most outstanding is that four families have been identified through Ancient DNA and Osteological Analysis. What our scientists have done for all of us, is to bring the Minoans to life in a way that has never been done before. Theirs are unique chapters, made possible by the existence of a unique site.

Appendix: catalogue

(original compilation by Vicky Kolivaki)
The order in which catalogue entries are presented below is in the same order as the tombs were discussed in the chapter (55, 89, 149, 198, 210, 203).

TOMB 55

Chronology: LM IIIB:1–IIIB:2

Contents of chamber

POTTERY
Amphoriskos
> R.M. 1830. Globular, decorated (wavy line FM.53). H.: 13.5 cm. Workshop: Armenoi. LM IIIB.

Cup
> R.M. 1831. Conical. H.: 3.5 cm. Workshop: Unknown. LM IIIB.

Stirrup jars
> R.M. 1826. Globular, decorated (elaborate triangles FM.71; lozenge FM.73). H.: 14 cm. Workshop: Kydonia. LM IIIB:1 end.
> R.M. 1827. Squat, decorated (octopus FM.21; sea anemone FM.27). H.: 11.5 cm. Workshop: Kydonia. LM IIIB:1 end.
> R.M. 1828. Squat, decorated (Minoan flower variant of FM.18C). H.: 8 cm. Workshop: Kydonia. LM IIIB:1.
> R.M. 1829. Globular, decorated (Minoan flower variant of FM.18C; sea anemone FM.27; zig-zag FM.61; adder mark FM.69). H.: 13 cm. Workshop: Kydonia. LM III B:1.

BRONZES
> R.M.M. 347. Razor, one-edged. Blade decorated with incised lines. Found with Skeleton 55Γ. L.: 22 cm. LM IIIB.
> R.M.M. 382. Knife. L.: 22.5 cm. LM IIIB.
> R.M.M. 384. Razor, one-edged. L.: 17.8 cm. LM IIIB.
> R.M.M. 345. Dagger. L.: 28 cm. Found with Skeleton 55Γ. LM IIIB.
> R.M.M. 392. Razor, one-edged. L.: 14.9 cm. LM IIIB.

R.M.M. 580. Knife one-edged with T-shaped pommel. Fragments of ivory hilt plates survive. Blade decorated with four incised lines. L.: 17.6 cm. LM IIIB.

SEALS

R.M. Σ45 (*CMS* V no. 273 cm). Amygdaloid. Hard green opaque stone: jasper (?). Edge damaged; face cracked. L.: 2.1 cm; W.: 1.8 cm. Sepia. MM III–LM I; 'talismanic' style.

R.M. Σ46 (*CMS* V no. 274). Lentoid. Serpentine. Reverse abraded. D.: 1.8 cm. Bird-lady; twig-like filler. LM.

LARNAKES

R.M. 1703. Larnax A. Gable lid. L: 0.94 m; W: 0.40 m; H: 0.75 m (with the lid). Body: first long side: octopus FM.21. Second long side: stemmed spiral FM.51; horizontal bands; horns of consecration FM.36 with double axe FM.35. First narrow side: papyrus FM.11; lozenge chain FM.73. Second narrow side: Minoan flower variant of FM.18; papyrus derivatives FM.11; hatched leaves. Lid: first and second long sides: alternating arcs FM.24. First and second narrow sides: undecorated. LM IIIB: 1.

R.M. 1846. Larnax B. Flat lid. L.: 0.695 m; W.: 0.405 m; H.: 0.658 m (with the lid). Body: first and second long sides: octopus tentacles FM.53:14. First narrow side: quatrefoil FM.55:4. Second narrow side: diagonal pattern FM.55:5. Lid: wavy band FM.53. LM IIIB: 2.

SMALL FINDS

Beads

R.M.Y. 539. Ca. 70 spherical, faience beads. D.: 0.2 cm. 3 empty gold covers survive. LM IIIB.

R.M.L. 2929. 9 beads. 3 spherical faience beads have incised lines. D.: 0.6–0.9 cm. 2 spherical faience beads have relief bands. D.: 0.8 cm. 2 faience beads are biconical. L.: 1.3 cm. 1 spherical rock crystal bead. D.: 0.7 cm. 1 spherical glass paste bead. D.: 0.8 cm. 1 barrel-shaped carnelian bead. L.: 1 cm. LM IIIB.

Jewellery

R.M.M. 394. 3 tube-shaped beads with remains of thread. L.: 0.7 cm. 2 spherical beads. D.: 0.3 cm and fragments of 1 more spherical bead. LM IIIB.

R.M.M. 410. Burial A. Ring of plate very corroded. D.: 1.5 cm; Th.: 0.4 cm. LM IIIB.

R.M.M. 418. Burial A. 2 fragments from a wire ring. D.: 1.7 cm; Th.: 0.3 cm. LM IIIB.

R.M.M. 419. Burial A. 2 fragments from a wire ring. Th.: 0.3 cm. LM IIIB.

R.M.M. 422. Under Larnax A. 3 fragments. Th.: 0.5 cm. LM IIIB.

R.M.M. 422b. Burial C. Fragments of 2 nails. L.: 1.3 cm. LM IIIB.

R.M.M. 427. Burial A. Bracelet of wire with coiled wire over the join. D.: 7 cm.; Th.: 0.3 cm. LM IIIB.

R.M.M. 3331. Silver ring. D.: 1.4 cm.

R.M.M. 3331b. Burial A. Lead fragments. L.: 5 cm.

Contents of dromos

SHERDS

Stirrup jar

R.M. 21125. 1 body sherd. Pres. Max. D.: 5 × 7.5 cm. LM IIIB:1.

TOMB 89

Chronology: LM IIIA:1–LM IIIA:2

Contents of chamber

POTTERY

Cups

R.M. 2145. Miniature, semi-spherical, undecorated. H.: 6.2 cm. Workshop: Unknown. LM IIIA.

R.M. 2146. Miniature, conical, undecorated. H.: 5.2 cm. Workshop: Unknown. LM IIIA.

Jugs

R.M. 2143. Decorated (traces of paint). H.: 10 cm. Workshop: Armenoi. LM IIIA.

R.M. 2144. Beak-spouted, decorated (iris FM.10A). H.: 8.5 cm. Workshop: Knossos. LM IIIA:1/ IIIA:2.

R.M. 2147. Miniature jug decorated (iris FM.10A). H.: 5 cm. Workshop: Knossos. LM IIIA:2.

R.M. 2148. Beak-spouted, decorated (concentric semicircles FM.43). H.: 8.5 cm. Workshop: Knossos. LM IIIA:2.

R.M. 2149. Beak-spouted, undecorated/grey in colour. H.: 6.3 cm. Workshop: Unknown. LM IIIA.

R.M. 2150. Miniature, beak-spouted, undecorated/grey in colour. H.L.: 5.7 cm. Workshop: Unknown. LM IIIA:2.

R.M. 2151. Miniature, bridge-spouted, handmade, undecorated/grey. H.: 6.5 cm. Workshop: Unknown. LM IIIA:2.

SMALL FINDS

Beads

R.M.Y. 554, L2936. Under Skeletons E–M. 16 beads in total. 2 ring-shaped, carnelian beads. L: 1.0 cm. 4 ring-shaped, faience beads have incised lines. D: 0.8 cm. 9 faience beads are biconical with incised lines/tree of them in fragments. L: 1.5 cm. 1 faience spherical bead. L: 1.2 cm. LM IIIA:2.

Conuli

R.M.L. 3036 a. One conical, reddish steatite conulus. H.: 2.00 cm. LM III.

R.M.L. 3036 b–d. 3 conuli in total. 1 conical, grey steatite, 1 conical, reddish steatite, 1 conical from whitish onyx. H.: 1.50–2.20 cm. LM III.

R.M.P. 23688 a. 1 conical clay conulus. H.: 1.80 cm. LM III.

R.M.P. 23688 b. 1 conical clay conulus. H.: 3.20 cm. LM III.

Contents of dromos
None

TOMB 149

Chronology: LM IIIA:2–LM IIIB:1

Contents of chamber

POTTERY
Alabastra

R.M. 3396. Cylindrical, 2 handles, decorated (concentric arcs FM.44). H.: 10.5 cm. Workshop: Knossos. LM IIIB:1.

R.M. 3422. Cylindrical, 3 handles, decorated (papyrus FM. 11). H.: 20.5 cm. Workshop: Armenoi. LM IIIA:2.

Jug

R.M. 3416. 1 handle, decorated (concentric bands FM.44). H.: 6.2 cm. Workshop: Armenoi. LM IIIB middle.

Stirrup jars

R.M. 3388. Squat-conical, new made, decorated (Mycenaean flowers FM.18C). Pristine condition. H.: 13.5 cm. Workshop: Kydonia. LM IIIB:1.

R.M. 3390. Globular, decorated (octopus FM.21). Pristine condition. H.: 15 cm. Workshop: Kydonia. LM IIIA:2.

R.M. 3407. Piriform, miniature, decorated (bands). H.: 13.8 cm. Workshop: Knossos. LM IIIA:2.

R.M. 3413. Squat-conical, miniature, decorated (scale pattern FM.70). Pristine condition. H.: 9 cm. Workshop: Kydonia. LM IIIA:2/IIIB:1.

R.M. 3418. Squat-conical, decorated (octopus FM.21). H.: 10.3 cm. Workshop: Knossos. LM IIIB:1 early.

R.M. 3421. Piriform, decorated (papyrus FM.11). H.: 14 cm. Workshop: Armenoi: Imitation of Knossos prototype. LM IIIB.

SHERDS
Bowl

R.M. 21113. One sherd: body and everted rim. Pres. max. dim.: Th.: 0.3 cm. LM IIIA.

BRONZES

R.M.M. 575. Spearhead; L.: 11.9 cm. LM IIIA:2.

SMALL FINDS
Beads

R.M.Y. 585a 1 spherical faience bead with relief lines. D.: 0.7 cm. LM IIIA.

R.M.Y. 585b 1 tube-shaped faience bead with incised lines. L.: 1.5 cm. LM IIIA.

Ivory

R.M.O. 1085. 2 fragments from ivory comb. L.: 3.4 cm. LM IIIA.

Contents of dromos
None

TOMB 198

Chronology: LM IIIA:2–LM IIIB:2

Contents of chamber

POTTERY
Alabastra

R.M. 6590. Cylindrical, 2 handles, decorated (foliate band FM.64). H.: 8.3 cm. Workshop: Kydonia. LM IIIB.

R.M. 6594. Baggy, 3 handles, decorated (net pattern FM.57). H.: 6.5 cm. Workshop: Unknown. LM IIIA:2.

R.M. 6595. Cylindrical, 3 handles, miniature, pristine, decorated (net pattern FM.57). H.: 5.6 cm. Workshop: Armenoi 2. LM IIIA:2/IIIB:1.

R.M. 6597. Baggy, 2 handles, miniature, decorated (foliate band FM.64). H.: 4.9 cm. Workshop: Knossos. LM IIIA:2/IIIB:1.

R.M. 6598. Baggy, miniature, new made, decorated (panelled pattern FM.75). H.: 6.1 cm. Workshop: Kydonia. LM IIIA:2.

Cups

R.M. 6588. Cylindrical, with spout, decorated (panelled patterns FM.75). H.: 7.3 cm. Workshop: Unknown. LM IIIB:1 beginning

R.M. 6589. Semi-spherical, miniature, decorated (foliate band FM.64). H.: 3.8 cm. Workshop: Armenoi. LM IIIB:2

Double vase

R.M. 6600. 2 jugs joined, miniature, with animal proteome on 1 handle; decorated (curved stripes FM.67). H.: 7.5 cm. Workshop: Armenoi. LM IIIA:2 (see Chapter 10 postscript).

Jars

R.M. 6591. Piriform, 3 handles, decorated (foliate band FM.64). H.: 18 cm. Workshop: Mycenaean import. LM IIIA:2.

R.M. 6592. Piriform, 3 handles, decorated (net pattern FM.57). H.: 16 cm. Workshop: Knossos, imitation of Mycenaean prototype. LM IIIA:2.

Jugs

R.M. 6585. Beak-spouted, decorated (V-shape FM.59). H.: 8.8 cm. Workshop: Armenoi. LM IIIA:2.

R.M. 6587. Decorated (concentric semicircles FM.43). H.: 6 cm. Workshop: Kydonia. LM IIIB:1.

R.M. 6596. Beak-spouted, decorated (foliate band FM.64). H.: 5.9 cm. Workshop: Kydonia. LM IIIA:2.

R.M. 6599. Beak-spouted, decorated (scale pattern FM.70). H.: 9.4 cm. Workshop: Armenoi, imitation of Knossos prototype. LM IIIA:2.

Stirrup jars

R.M. 6586. Squat-conical, new made, decorated (palm trees FM.15). H.: 13 cm. Workshop: Armenoi. LM IIIA:2.

R.M. 6593. Squat, miniature, new made, decorated (Mycenaean flowers FM.18A). H.: 6.1 cm. Workshop: Kydonia. LM IIIA:2.

SHERDS

Bowl

R.M. 21110. 2 body sherds, decorated (net pattern). Pres. max. dim.: Th.: 0.2 cm. LM IIIA:2.

BRONZES

R.M.M. 610. Razor, leaf-shaped. L.: 17.2 cm. LM III:A/ IIIB:1.

R.M.M. 611. Razor, leaf-shaped. L.: 19.3 cm. LM III:A/ IIIB:1.

SMALL FINDS

Jewellery

R.M.M. 630. 1 wire ring, snake imitation. D.: 1.8 cm. LM III.

R.M.M. 679. 1 wire ring, snake imitation. D.: 1.7 cm. LM III.

R.M.M. 680a. Burial A. 5 fragments from a ring of wire D.: 1.8 cm. LM III.

R.M.M. 680b. 1 wire ring. D.: 1.7 cm. LM III.

Four fragments from a ring of wire. Th.: 0.1 cm.

R.M.M. 681a. 2 fragments from a ring. D.: 2.40 cm. LM III.

R.M.M. 682. 1 tube-shaped bead, very corroded. L.: 0.8 cm. LM III.

R.M.M. 682b. 1 cylindrical bead. L.: 0.4 cm. LM III.

Beads

R.M.Y. 622. Vase 8. 1 tube-shaped faience bead with incised lines. L.: 1.5 cm. LM IIIA.

R.M.Y. 622a. Inside vase 9. One biconical faience bead with incised lines and fragment of another one. L.: 1.5 cm. LM IIIA.

R.M.Y. 622b. 11 beads in total. 3 biconical faience beads with incised lines. L.: 1.5 cm. 3 tube-shaped faience beads. L.: 1.5 cm. 3 spherical faience beads. D.: 0.6 cm. 2 fragments of 1 oval-shaped faience bead. Fragment from bone clevis. L.: 1.5 cm. LM IIIA.

Conuli

R.M.L. 3051 a–b. 2 conuli. 1 conical, grey steatite conulus. H.: 1.80 cm. 1 conical, miniature (for child?) steatite conulus. H.: 0.80 cm. LM III.

R.M.L. 3051 c. 1 conical, grey steatite conulus. H.: 1.20 cm. LM III

R.M.P. 23697 a. 1 conical, clay conulus. Black slip outside. H.: 2.00 cm. LM III.

R.M.P. 23697 c. 1 conical, clay conulus. Black slip outside. H.: 2.30 cm. LM III.

NICHES

North niche: no finds

South niche: bone fibula and a bead (information from notebooks; not extant).

CONTENTS OF DROMOS

Beads

R.M.L. 2996. Niche. 1 bone fibula with relief head. L.: 3.5 cm.

Five ring-shaped bone beads. D.: 0.3–0.8 cm. 1 biconical carnelian bead. L.: 2.0 cm. LM IIIA.

TOMB 210

Chronology: LM IIIA:1/LM IIIA:2–LM IIIB:1 late

Contents of chamber

POTTERY

Alabastron

R.M. 7673. Globular, 3 handles, miniature, decorated (Mycenaean flowers FM.18A). H.: 7.8 cm. Workshop: Knossos. LM IIIA:1–LM IIIA:2.

Cup

R.M. 7665. Semi-spherical, miniature, decorated (fully painted). H.: 3.3 cm. Workshop: Knossos. LM IIIA:2–LM IIIB:1

Jar

R.M. 7663. Piriform, 3 handles, decorated (net pattern FM.57). H.: 11.3 cm. Workshop: Armenoi. LM IIIA:2.

Jugs

R.M. 7664. Bridge-spouted, decorated (net pattern FM.57). H.: 6.1 cm. Workshop: Unknown. LM IIIA:2–LM IIIB:1.

R.M. 7666. Decorated (foliate band FM.64). H.: 5.5 cm. Workshop: Armenoi. LM IIIB:1 late.

R.M. 7667. Beak-spouted, miniature, decorated (adder mark FM.69). H.: 5.8 cm. Workshop: Armenoi. LM IIIB:1 beginning.

R.M. 7668. Beak-spouted, decorated (quirk FM.48). H.: 8.5 cm. Workshop: Armenoi. LM IIIB.

R.M. 7672. Beak-spouted, decorated (quirk FM.48, foliate band FM.64). H.: 7.6 cm. Workshop: Armenoi. LM IIIA:2.

Stirrup jar

R.M. 7669. Piriform, decorated (elaborate triangles FM.71). H.: 12 cm. Workshop: Knossos. LM IIIB:1.

SHERDS

Jug

R.M. 21066. ca. 16 sherds. Base, wall, spout and 1 vertical handle. Workshop: Unknown. LM IIIA:2.

SMALL FINDS

Beads

R.M.Y. 633a. 2 spherical faience beads with incised lines. D.: 0.8 cm, 1.2 cm. LM IIIA.

R.M.Y. 633b. 4 spherical faience beads. D.: 0.8 cm. LM IIIA.

R.M.Y. 633c. 1 tube-shaped calcite bead with cloth remains. L.: 3.0 cm. LM IIIA.

R.M.Y. 633d. 1 cylindrical faience bead. L.: 3.0 cm. LM IIIA.

R.M.L. 3003. 6 spherical cornaline beads. D.: 0.7 cm. LM IIIA.

Conuli

R.M.L. 3065 a–c. 3 conical, redish steatite conuli. H.:1.00–2.00 cm. LM III.

R.M.L. 3065 d–e. 2 conical conuli, 1 from reddish steatite and 1 from grey steatite. H.:1.00–2.00 cm. LM III.

Contents of dromos

Jug

R.M. 20863. Ca. 16 sherds: wall, vertical strap handle and everted rim. Pres. max. dim.: wall th: 0.6 cm. LM IIIA.

TOMB 203

Chronology: LM IIIA:1/LMIII A:2–LM IIIB:1

Contents of chamber

POTTERY

Alabastra

R.M. 7516. Globular, decorated (quirk FM.48). H.: 8.1 cm. Workshop: Armenoi, imitation of Knossos prototype. LM IIIA:2.

R.M. 7517. Globular, new made, decorated (papyrus FM.11). H.: 9 cm. Workshop: Knossos, imitation of Mycenaean prototype. LM IIIA:1–LM IIIA:2.

R.M. 7521. Cylindrical, 3 handles, decorated (lozenge FM.73). H.: 13.1 cm. Workshop: Kydonia. LM IIIB:1.

R.M. 7522. Squat, 3 handles, decorated (traces of paint). H.: 5.7 cm. Workshop: Unknown. LM IIIA:2.

R.M. 7526. Globular, decorated (stipple pattern FM.77). H.: 10.3 cm. Workshop: Knossos. LM IIIA:1.

Cup

R.M. 7519. Semi-spherical decorated (concentric semicircles FM.43). H.: 6.5 cm. Workshop: Armenoi 2. LM IIIA:2.

Jars

R.M. 7518. Piriform, 3 handles, decorated (net pattern FM.57). H.: 15 cm. Workshop: Knossos, imitation of Mycenaean prototype. LM IIIA:2.

R.M. 7523. Piriform, 3 handles, decorated (foliate band FM.64). H.: 18.2 cm. Workshop: Knossos, imitation of Mycenaean prototype. LM IIIA:2.

Jugs

R.M. 7520. Beak-spouted, decorated (concentric semicircles FM.43). H.: 8.3 cm. Workshop: Armenoi. LM IIIB:1.

R.M. 7524. Beak-spouted, decorated (traces of paint). H.: 7.8 cm. Workshop: Unknown. LM IIIA:2.

Unidentified sherds found in the northern area of the chamber.

SHERDS

Jugs

R.M. 21112. 2 sherds: wall and vertical handle. Pres. max. dim.: wall th.: 0.2 cm. LM IIIA. Workshop: Kydonia.

R.M. 21114. 5 body sherds. Decorated (concentric arches). Pres. max. dim.: wall th: 0.4 cm. LM IIIA.

BRONZES

R.M.M. 591. Razor, leaf-shaped. L.: 21.8 cm. LM IIIA–LM IIIB:1.

Fragments of rings or earrings.

SEALS

R.M. Σ163. (*CMS* V Suppl. 1B no. 312). Lentoid. Chlorite or serpentine. Heavily abraded; edges battered. D.: 1.72–1.75 cm. Late Minoan.

SMALL FINDS

Jewellery

R.M.M. 631. 4 fragments from a wire ring D.: 1.90 cm. LM III.

R.M.M. 688. 3 fragments from two wire rings D.: 1.70 cm. LM III.

R.M.M. 689. 5 cylindrical beads. L.: 0.6 cm. LM III.

R.M.M. 690. Two-thirds of a wire ring D.: 1.80 cm. LM III.

R.M.M. 691. 3 fragments in total, 1 is part of a ring Th.: 0.40 cm. LM III.

R.M.M. 692a–b. 16 fragments from 3 wire rings Th.: 0.1 cm. LM III.

R.M.M. 693. Two-thirds of a wire ring. D.: 1.70 cm. LM III.

R.M.M. 694a. 1 spherical bead. D.: 0.3 cm. LM III.

R.M.M. 694b. 1 spherical bead. D.: 0.4 cm. LM III.

R.M.M. 695a–b. 8 fragments from a ring. Th.: 0.40 cm. LM III.

Beads

R.M.L. 2999, Y625. 26 beads in total. 22 spherical cornaline beads. D.: 0.6 cm. 3 tube-shaped faience beads. L.: 0.7 cm. 1 biconical faience bead. L.: 1.5 cm. LM IIIA.

R.M.L 2999. 61 beads in total. 58 spherical cornaline beads. D.: 0.6 cm. 1 tube-shaped cornaline bead. L.: 1.5 cm. 1 biconical cornaline bead. L.: 1.8 cm. 1 spherical rock crystal bead. D.: 1.0 cm. LM IIIA.

Conuli

R.M.L. 3061 a–b. 2 conical, grey steatite conuli. H.: 1.00–1.20 cm. LM III.

R.M.L. 3061 c. 3 conical, light grey steatite conuli. H.:1.00–1.20 cm. LM III.

R.M.P. 23707 a. 1 conical and fragment of a second one, reddish clay conulus. Slip outside. H.: 2.00 cm. LM III.

R.M.P. 23707 b. 1 conical, clay conulus. H.: 1.80 cm. LM III.

R.M.P. 23707 c. Fragment of conical, brown clay conulus. Slip outside. H.: 2.20 cm. LM III.

Contents of dromos

POTTERY

Kylikes

R.M. 21101. 1 sherd: stem. Pres. max. dim.: H.: 6.5 cm. LM IIIA:2.

Bibliography

Buikstra, J.E. and Ubelaker, D.H. (1994) *Standards for Data Collection from Human Skeletal Remains.* Fayetteville AR, Arkansas Archaeological Survey.

Driessen, J. (1990) *An Early Destruction in the Mycenaean Palace at Knossos: A New Interpretation of the Excavation Field-Notes of the South-East Area of the West Wing.* Leuven, *Acta Lovaniensia* Monographiae 2.

Godart, L. and Tzedakis, Y. (1992) *Témoignages archéologiques et épigraphiques en Crète occidentale du Néolithique au Minoen Récent III B,* Rome, Incunabula Graeca 93.

Lazaridis, I., Mittnik, A., Patterson, N., Mallick, S., Rohland, N., Pfrengle, S., Furtwängler, A., Peltzer, A., Posth, C., Vasilakis, A., McGeorge, P.J.P., Konsolaki-Yannopoulou, E., Korres, G., Martlew, H., Michalodimitrakis, M., Özsait, M., Özsait, N., Papathanasiou, A., Richards, M., Roodenberg, S.A., Tzedakis, Y., Arnott, R., Fernandes, D.M., Hughey, J.R., Lotakis, D.M., Navas, P.A., Maniatis, Y., Stamatoyannopoulos, J.A., Stewardson, K., Stockhammer, P., Pinhasi, R., Reich, D., Krause, J. and Stamatoyannopoulos G. (2017) Genetic origins of the Minoans and Mycenaeans. *Nature* 548, 214–18. [https://doi.org/10.1038/nature23310]

Richards, M.P. and Hedges, R.E.M. (2008) Stable isotope results from the sites of Gerani, Armenoi and Mycenae. In Y. Tzedakis, H. Martlew and M.K. Jones (eds), *Archaeology Meets Science: Biomolecular and Site Investigations in Bronze Age Greece,* 220–30. Oxford, Oxbow Books.

Richards, M.P., Smith, C., Nehlich, O., Grimes, V., Weston, D., Mittnik, A., Krause, J., Dobney, K., Tzedakis, Y. and Martlew, H. (2022) Finding Mycenaeans in Minoan Crete? Isotope and DNA analysis of human mobility in Bronze Age Crete. *PLoS One* 17(8), e0272144. [https://doi.org/10.1371/journal.pone.0272144]

Weston, D.A. (2019) Human osteology', In M.P. Richards and K. Britton (eds), *Archaeological Science: An Introduction,* 169–74. Cambridge, Cambridge University Press.

Part II

The 'City' of Armenoi

The identification of *da-*22-to*

Louis Godart

The spectacular finds from the Necropolis and site of Armenoi mean that this place certainly had an essential role in Late Minoan III Crete. We should therefore expect that the Minoan name of the city of Armenoi appears in the lists of toponyms that are given to us on the tablets of Knossos. But among the tablets in Linear B that appear in the archives of Knossos, only a few toponyms correspond to names attested today as *ko-no-so* =Κνωσός, *ku-do-ni-ja* = Κυδωνία, *a-pa-ta-wa* = Ἄπτερα, *pa-i-to* = Φαιστός, the others have names extinct since the Mycenaean age.

What could have been the name of the site that is now called Armenoi in the texts from the palace of Minos? A series of elements leads us to retain that the toponym *da-*22-to* is likely to designate the site of Armenoi:

1) the analysis of the groupings of placenames mentioned in the texts of Knossos makes it possible to locate *da-*22-to* in the region between the plain of the Messara and the valley of Amari;
2) the discovery in Eleusis of a stirrup jar bearing the name of *da-*22-to*, sent under the aegis of a Mycenaean 'king' (*wa* = abbreviation of the adjective *wa-na-ka-te-ro* = ἀνάκτερος) means that this city had a king, which fits very well with the royal character of the Necropolis;
3) this city of *da-*22-to* which exported stirrup jars to Knossos and to the continent had a port as indicated in the tablets of Knossos V 756 and V 1002.

It therefore appears that the name used by the Mycenaean scribes to designate the site of Armenoi was *da-*22-to*.

Before attempting to associate the localities of Armenoi, Vrysinas, Pigi, Maroulas and Chamalevri, with toponyms attested in the tablets of Knossos, let us recall the names of towns used by the scribes of Knossos, which referred to localities as 'West Cretan'. It has long been clear that the six toponyms present in the tablets of the **Co** Knossos series, written by the scribe 107, all refer to localities in western Crete (Godart 1972; Godart and Tzedakis 1992, 222–59). These are the toponyms *ku-do-ni-ja*, *a-pa-ta-wa*, *si-ra-ro*, *ka-ta-ra-i*, *wa-to* and *o-du-ru-we*. Among these six locality names, two of them correspond to toponyms still in use today: *ku-do-ni-ja* (Κυδωνία) and *a-pa-ta-wa* (Ἄπτερα). It is also clear that, in addition, the toponyms and toponymic derivatives painted on the bodies of the stirrup jars discovered at Thebes, Eleusis, Mycenae, Tiryns and Midea relate to towns and individuals of western Crete because the clay of these vases is a clay from that region as demonstrated by scientific analysis (Catling *et al.* 1980).

These are the following localities:]*e-ra* in **MY Z 202**, **56-ko-we* in **TI Z 27**, *da-*22-to* in **EL Z 1**, *o-du-ru-wi-jo* (ethnic masculine of *o-du-ru-we*) in **TH Z 839**, *wa-to* in **TH Z 846, 849, 851, 852, 853, 854** and **882** as well as *si-ra-]ri-jo* in **TI Z 29**. Let us add that the ethnic (see below) *wa-ti-jo* is present in the tablet **KH Ar 2**.1 and the ethnic *pu-na-si-jo* in **KH Ar 2**.2. In the Linear B tablets the term 'ethnic' means 'men or women from a specific locality', for example: *ko-no-si-jo* means 'The man [or the men] from Knossos'; *ko-no-si-ja* 'The woman [or the women] from Knossos'.

To these toponyms and toponymic derivatives which evoke localities in western Crete we must therefore add the town of *pu-na-so* because it is attested in tablet **KH Ar 2.2** which was discovered in Chania (Godart and Tzedakis 1992, 188–9). This tablet allows us to expand the list of Cretan cities to be associated with West Crete because in one of the Knossos tablets (**C 979**) we find a provincial leader with the title of *a-to-mo* associated with four closely related localities and among them *pu-na-so: do-ti-ja, ra-ja, pu-na-so-qe, ra-su-to-qe*. This person is undoubtedly the representative of the central power in these four cities and as the price of his labour he receives a pig:

C 979

do-ti-ja / ra-ja, 'pu-na-so-qe, ra-su-to-qe' a-to-mo SUS 1

Since *pu-na-so* is in West Crete, as it appears from the tablet **KH Ar 2**, it is obvious that the three cities associated with *pu-na-so* in **KN C 979**, and under the control of the official *a-to-mo*, must be also located in western Crete. It is therefore appropriate to add the towns of *do-ti-ja, ra-ja* and *ra-su-to* to the list of Cretan towns mentioned in the Linear B tablets of Knossos which relate to western Crete.

That gives us a total of 13 cities: *a-pa-ta-wa, da-*22-to, do-ti-ja, e-ra, ka-ta-ra-i, ku-do-ni-ja, o-du-ru-we / o-du-ru-wi-jo, pu-na-so / pu-na-si-jo, ra-ja, ra-su-to, si-ra-ro / si -ra-ri-jo, wa-to / wa-ti-jo, * 56-ko-we*.

Some of these cities are part of a block that we could define as 'the block of the Cretan far west', the cities of *a-pa-ta-wa, ka-ta-ra-i, ku-do-ni-ja, o-du-ru-we, si-ra-ro* and *wa-to* cited in the tablets **Co** of the scribe 107. They have the common characteristic of being centres of breeding of small and large cattle, which require the presence of fertile and well irrigated land. Two of these localities, *ku-do-ni-ja* and *a-pa-ta-wa* (Κυδωνία, Ἄπτερα) correspond to the cities of Chania and Aptera, whose territory have these characteristics. It is therefore in the fertile region located between the modern city of Chania and the territories around Souda Bay that we will locate the other cities of the **Co** series (*ka-ta-ra-i, o-du-ru-we, si-ra-ro* and *wa-to*).

Where can we locate the other cities that we have just identified (*da-*22-to, do-ti-ja, e-ra, pu-na-so / pu-na-si-jo, ra-ja, ra -su-to, * 56-ko-we*)? Systematically reproducing all the attestations of the *BOS, BOSᵐ, BOSᶠ, BOSˣ* logograms in the Knossos tablets allows us to understand the essential role played by West Crete in the breeding of bovids (Olivier *et al.* 1973, 331). The breeding of large livestock such as cattle requires well-irrigated land. This peculiarity is typical of the area of West Crete. The herds of cows (*BOSf*) are attested only in the tablets of the scribe 107. The texts **Co** apart, there are in fact bovids (male and female alike) in the tablets of the scribe 107: *39 BOSᶠ, 99 BOSᵐ* and *1 ta BOS* in texts **C 901, 989, 5544** and **5733**, which is much more than anything that can be found of bovids in the tablets of other scribes of Knossos. In fact, if we add to these 140 cattle the cows and bulls of **Co**, we obtain a total of some 180 head of cattle (*10 BOSᵐ* and *40 BOSᶠ* being returned for **Co**).

This critical role played by the West Cretans in the breeding of cattle is also apparent from a review of documents from the Room of Chariot Tablets. It should also be noted that the text **Ce 59** belonging to this group of tablets attributed by Jan Driessen to Late Minoan II, in turn proves that West Crete, at the time of the Mycenaean conquest of the island, already had the particularity of being a land especially suitable for raising large cattle.

The toponyms found in this document are in fact associated with bovid records.

Ce 59

.1 *]ma-sa we-ka-ta BOSm 6 // da-wo / we-ka-ta BOSm 6*
.2a *ta-ra-me-to [.]-mo*
.2b *ku-]ta-to / we-ka-ta BOSm 10 // da-*22-to / we-ka-ta BOS 6*
.3a *[.]-mo*
.3b *] tu-ri-so / we-ka-ta BOSm 6 //ku-do-ni-ja / we-ka-ta BOSm 50*

It is remarkable to note that the scribe '124c', the author of this document, as mentioned by Olivier in the *Les Scribes de Cnossos* (Olivier 1967, 66–76), records localities which belong to very precise regions of Crete, which extend from Tylissos to Chania.

The toponym] *ma-sa*, is attested in **Dq 42**.b, **Ga 1058, X 7776, X 5737**. The locative *ma-sa-de* (which, with the suffix *-δε*, means 'towards the town of *ma-sa*') in **X 744** and the restitution *ma-sa[-de* in **F 866**. The document **Ga 1058** tells us that in the locality of *ma-sa* there was a *te-o-po-ri-ja*, that is to say a 'procession' (θεhοφόρια; cf. θεοφόρος). This means that in the locality of *ma-sa* or in its surroundings, there was a sanctuary, a hypothesis that confirms the association between *di-ka-ta-de* (towards the sanctuary of Δίκτη) and *ma-sa[-de* (towards the sanctuary of Masa) in document **F 866**. It should be noted in passing that another 'procession' which concerned a locality in West Crete is mentioned in document **Od 696** because it is permissible to suppose, as imagined by Michel Lejeune, that the word *o-du-we* associated in this document with *te-o-po-ri-ja* is a graphic slip for *o-du <-ru> -we*, the locality of West Crete associated with *ku-do-ni-ja* (Κυδωνία) and *a-pa-ta-wa* Ἄπτερα) in the tablets of the scribe 107 and mentioned in the inscription on stirrup vessels **TH Z 839**.

In this document **Ce 59** the locality of *ma-sa* is associated with *da-wo*, a town located in the plain of Messara and close to Phaistos. It is logical to assume a geographical relationship between the two toponyms. In line .2 of the tablet the cities of *ku-ta-to* and *da-*22-to* are associated, while in line .3 we find *tu-ri-so*, modern Τυλισσός/Tylissós, on the one hand, and *ku-do-ni-ja* modern Κυδωνία/Chania on the other.

Document **Ce 59** therefore records oxen in an area that stretches from the foot of Psiloritis to the plain of Messara and the region of Chania. Hence it is sensible to locate the towns of *ku-ta-to* and *da-*22-to* in the region between Messara and Chania and imagine that their territories extend from the valley of Amari to the area of Rethymnon. It is therefore also possible to imagine that the towns of *da-*22-to, do-ti-ja, e-ko-so, e-ra, ku-ta-to, pu-na-so, ra-ja* and **56-ko-we* are to be located in the region mentioned above.

The series of documents written by the scribe 125 (series V, documents **V 756, 1002, 1003, 1004, 1005, 1043, 1583** and series X, documents **X 7577, 7670, 7797?, 7974**, relate in part to western Crete because we find in these texts, among others place-names, *a-pa-ta-wa* and *da-*22-to*.

The documents **V 756** and **V 1002** are complete and as such, make it possible to interpret the series. Let us take, for instance, **V 1002**:

V 1002

.A *'po-ti-ro', pi-ra-ki-jo 1 pe-ri-jo-ta-qe 1*
.B *da-*22- ti-ja* /

All the records of this scribe are built on the model of **V 1002** with a first word which is an ethnic feminine singular followed by the term *po-ti-ro* which precedes two anthroponyms associated with each other by the enclitic *-qe* = -τε. Among the ethnic groups we find *a-pa-ta-wa-ja* (**V 7670**) and *da-*22-ti-ja* (**V 756** and **1002**), as well as other localities named *ka-di-ti-ja* (**V 1003** and **V 9320**), *ki-ra-di-ja* (**V 1005**), *ku-pa-si-ja* (**V 1043**) and *di-pi-ja* (**V 7577**).

The word *po-ti-ro* undoubtedly serves to indicate the role or function of the two anthroponyms which follow. It appears that the only satisfactory explanation for this term is the Greek ποντίλος 'sailor', a synonym of the word ναυτίλος as illustrated in Aristotle (Liddel and Scott 1966, 1448, *s.v.* ποντίλος). These tablets therefore deal with transactions that concern the sea.

The feminine ethnic that opens each of the tablets in the series written by the scribe 125 cannot naturally relate to the male names that appear in each of the texts. It will therefore be necessary to find an explanation for this feminine.

What about the other localities mentioned in the series? The word *ka-di-ti-ja* recalls Κάδιστον ὄρος/, a locality on the northern coast of Crete; *ki-ra-di-ja* Σκιράδιον ἄκρον, a promontory of the island of Salamis where there is a temple dedicated to Athena Skiras and *ku-pa-si-ja* which suggests an adjective derived from the name Κύπασις which is that of a town of the Hellespont. *A-pa-ta-wa-ja* is, of course, an ethnic (see above) derivative of *a-pa-ta-wa* whose Greek correspondent is Ἄπτερα (Aptera), the famous city of West Crete near Souda Bay.

It is obviously not a coincidence that these names of cities evoke coastal localities and, ultimately, we have good reason to imagine that the last ethnic *da-*22-ti-ja*, brings to mind the city called *da-*22-to*, in turn a coastal locality or at least a city located not far from the coast. Therefore, the analysis of the documents of the scribe 125 of Knossos leads us to the conclusion that all the toponyms of set **V** (5) have to do with coasts and the sea, and if so, the interpretation of *po-ti-ro* as ποντίλος 'marine' emerges singularly strengthened.

However, it remains to be explained why the toponymic derivatives attested at the beginning of the tablet are in the

feminine singular. And on this subject the hypothesis of John Chadwick is attractive (Chadwick 1973, 199–201). We can, indeed, according to the English scholar, imply a word like (ναῦς) following each of these ethnic groups and thus consider that these terms serve to indicate the names of the home ports of these sailors or, at least, the names of the boats on which they served. Chadwick points out (Chadwick 1973, 199–201) that the first of these two hypotheses is the more credible since it can explain the two records of **V 756** and **V 1002** relating to *da-*22-to*. These two documents bear witness to different names because two different ships are called to leave this port on the northern coast of Crete.

The city of *da-*22-to* is attested in one of the vases with painted inscriptions found on the continent since in the text from Eleusis, we find the expression:

EL Z 1

.1 *da-*22-to*
.2 *da-pu2-ra-zo, wa,*

We have seen that this town is to be located in the region west of Psiloritis and, in particular, in the area between the valley of Amari and the outskirts of Rethymnon. Since a vase coming from *da-*22-to* is found on the Greek mainland, in Eleusis, one can think that this shipment left from the locality of *da-*22-to* very close to the sea. Therefore *da-*22-to*, a locality close to the sea since it controls ships and sailors, is to be placed in the vicinity of Rethymnon.

In this **EL Z 1** inscription, following the anthroponym *da-pu₂-ra-zo*, the abbreviation *wa* is attested which, of course, is an abbreviation of the adjective *wa-na-ka-te-ro* (ἀνάκτερος) the meaning of which is well established: 'belonging to the king'. The export of the product sailed in the stirrup jar connected with this *da-*22-to* and imported to the Greek mainland and, in this case to Eleusis in particular, is therefore placed under the aegis of a palatial authority.

The question that we should ask ourselves is the following: what is the palatial site near Rethymnon and what is the palatial authority that could have presided over this export? The **EL Z 1** stirrup jar dates from Late Minoan IIIB1, which means that it is after the fall of the palace of Knossos if one places, as most specialists agree, the latter in Late Minoan IIIA2. Therefore, the 'king' mentioned in the inscription discovered at Eleusis can in no way be the king (wanax = *wa-na-ka* ἄναξ) of Knossos.

It appears that the political rise of Kudonia (Chania) followed the fall of the palace of Knossos. Linear B is written in Chania and the Mycenaean administrators of *ku-do-ni-ja* manage the kingdom of West Crete according to the standards and customs of the administrators of all Mycenaean palatial centers. A wanax is in place in Chania as taught by the attestations of *wa* and *wa-na-ka-te-ro* present in the inscriptions in Linear B painted on the bodies of the stirrup jars from the localities that we have placed in far West Cretan (**Co** tablets, scribe 107.) It is this wanax of Kydonia which manages, among others, the cities of *wa-to* and *pu-na-so* that we find in the tablet **KH Ar 2** as well as the cities of *o-du-ru-we* and *si-ra-ro* whose toponymic derivatives we find in the stirrup jars discovered in Thebes and Tiryns (**TH Z 839** and **TI Z 29**).

But if Chania, ancient Kydonia, became an important palatial centre in the aftermath of the fall of Knossos, other sites in Crete and, in particular, in western Crete seem to be animated with significant vitality in the period after the end of the palace of Minos. One thinks in particular of the site of Armenoi, where excavation of the prodigious Necropolis attests to its importance and splendour. Some of the 232 tombs excavated are undoubtedly princely tombs and, as discussed elsewhere in this volume, Tomb 159 with its dromos of 15.50 m, staircase carved with 25 steps, monumental door, burial chamber and crypt, has all the appearances of a royal tomb.

In the Late Minoan III Necropolis of Armenoi, the excavation of Tomb 146 revealed a stirrup jar with a Linear B inscription which reads *wi-na-jo* (Fig. 9.39). It is remarkable that this inscription is in the same hand as those painted on the stirrup jars which also bear the word *wi-na-jo* discovered at two other sites, the Unexplored Mansion at Knossos on Crete, and Midea on mainland Greece, in the earth of Argolid. It is therefore more than possible that the stirrup jar in Tomb 146 at Armenoi, with the inscription *wi-na-jo*, was painted in the palatial town of Armenoi and that the two other stirrup jars with *Do*, with the anthroponym *wi-na-jo*, found in Knossos and Midea were exported from Armenoi to those cities. In addition, the presence in the Armenoi Necropolis of tin-plated vases, undoubtedly pleads in favour of trade between the 'city' of Armenoi and other localities in the eastern Mediterranean, as tin plated vessels are thought to have been exports from the Greek Mainland.

Let us try to summarise the elements of the puzzle:

1) The locality of *da-*22-to* at the time of the splendour of Knossos was particularly important as documented by the Linear B tablets discovered in the palace of Minos. For *da-*22-to* there are records of personnel (craftsmen), livestock, agricultural products, spices, fabrics and land. The *da-*22-to* workshops, where furniture and precious objects as taught by the tablets of the scribe 102 were made, were under the direction of various βασιλεῖς (basileis) (**As 40**) (Godart 2020, 284–94). The function of the βασιλεύς, Mycenaean *qa-si-re-u* in the singular, *qa-si-re-we*/βασιλεῖς in the plural, is always linked to the craft activity (the manufacture of furniture in Knossos and Pylos, the processing of bronze in Pylos, the production of vases in Knossos and pieces of armament in Thebes); the *qa-si-re-u* could be called to take on the

territory where he exercised his functions, the goods to be sent to the palace (for example the gold supplied in **Jo 438** of Pylos by the *qa-si-re-u a-ke-ro*, and the oxen skins delivered by the various Theban *basileis* in **Uq 434**); in his capacity as responsible for a *qa-si-re-wi-ja* the workshop where a basileus was the chief), the *qa-si-re-u* could count on the collaboration of other people.

Between the 2nd millennium and the 9th century BC, which saw Homer write the *Iliad*, the terms ἄναξ and βασιλεύς have undergone a remarkable semantic evolution.

The political structures to which the *Iliad* but above all the *Odyssey* allude, date back to the 1st millennium. The term βασιλεύς, in Homer, indicates the king invested with absolute authority. The term has, in Homeric poems, a political meaning while ἄναξ has become only a title that can be applied to anything (a horse, sleep, or Agamemnon, *etc*): anyone can be called ἄναξ while there is only one βασιλεύς.

In Linear texts B on the other hand, ἄναξ is the real function name. The *wa-na-ka* of the tablets is the Head of State: he possesses all the prerogatives related to this office since he is head of the administration, of the army, lord of the palace, responsible, in part at least, for the management of the sanctuaries and above all, there is only one ἄναξ in each State.

The basileis, *qa-si-re-we* in the plural, are, on the contrary, simple officials of provincial rank responsible for the workshop of a potter, carpenter or the workshop of a blacksmith and there are many basileis in every Mycenaean State. Between the Mycenaean world and the Homeric Age, the semantic value of the two terms changes radically: the head of the Mycenaean State was defined as ἄναξ (a term devalued in Homer), while the word βασιλεύς used in the Mycenaean tablets to indicate the person responsible for a group of craftsmen, designates in the *Iliad* and the *Odyssey*, the Head of State invested with all the prerogatives related to the office.

The fact of finding basileis in *da-*22-to* means that there were various workshops of artisans in the city, headed by a βασιλεύς. In the Necropolis of Armenoi, the discovery of numerous high quality bronze tools and weapons implies the existence in the city connected with the Necropolis of important workshops where metals were worked. The attestation of various basileis associated with *da-*22-to* is a confirmation of the identification of the Mycenaean toponym with modern Armenoi.

2) This locality can be placed in the region which extends from Messara and the valley of Amari and leads to the northern coast of Crete.

3) The locality of *da-*22-to* controlled ships as the tablets of the scribe 125 of Knossos attest.

4) Consequently, *da-*22-to* appears to be located in the vicinity of Rethymnon and relatively close to the sea.

5) The city of *da-*22-to* = Armenoi exported stirrup jars to Eleusis but also to Knossos and Midea.

Thus Armenoi has been in contact with Knossos, the Greek Mainland and the eastern Mediterranean as it appears also thanks to the presence of imported tin-plated kylikes (goblets) found in what is thought to be a royal tomb (159) at the Necropolis.

6) At least some of these exports took place under the aegis of a wanax as shown by the inscription **EL Z 1** where we read the abbreviation *wa* = *wa-na-ka-te-ro* 'belonging to the king'.

7) Tomb 159 has all the appearances of a royal tomb. That fits very well with the mention of a king (*wa* = *wa-na-ka-te-ro* 'belonging to the king') on the inscription **EL Z 1**.

Therefore, we have every reason to believe that the toponym *da-*22-to* in Linear B was used to indicate the site of the town of Armenoi as it existed during the time of the Mycenaean presence at Knossos.

Bibliography

Catling, H.W., Cherry J.F., Jones, R.E. and Killen, J.T. (1980) The Linear B inscribed stirrup-jars and western Crete. *Annual of the British School at Athens* 75, 49–113.

Chadwick J. (1973) A Cretan Fleet? *Antichità Cretesi*, vol. I, Catania, 199–201.

Godart, L. (1972) Les tablettes de la série Co de Cnossos: Acta Mycenaea II. *Minos* 12, 418–24.

Godart, L. (2020) *Da Minosse a Omero*. Turin, Einaudi.

Godart, L. and Tzedakis, Y. (1992) *Témoignages archéologiques et épigraphiques en Crète occidentale du Néolithique au Minoen Récent III B*. Rome, Incunabula Graeca 93.

Liddel H.D. and Scott, R. (1966) *A Greek–English Lexicon*, Oxford, Oxford University Press.

Olivier, J.-P. (1967) *Les scribes de Cnossos*. Rome, Incunabula Graeca 17.

Olivier, J.-P., Godart, L., Seydel, C. and Sourvinou, C. (1973) *Index généraux du linéaire B*. Rome, Incunabula Graeca 52.

Studies at the 'city' of Armenoi, the Necropolis and their environs

Peter W. Ditchfield and Holley Martlew

Studies of collagen carbon and stable isotope values from faunal remains recovered from the dromos of Tomb 159 and of the geology and minerals identified in the region around the 'city' and Necropolis provide some indication of husbandry practices, sources of materials used in the manufacture of objects recovered and aspects of tomb construction.

Faunal collagen carbon and nitrogen stable isotopic analysis

Peter W. Ditchfield

Twelve faunal bones from the dromos of Tomb 159 and 20 faunal bones from the nearby town site were selected for carbon and nitrogen stable isotopic analysis (see Tables 5.4 and 5.5 above for taxa and skeletal elements represented). Bone collagen was extracted from the faunal bones following the modified Longin (1971) method. Carbon and nitrogen stable isotopic analyses were undertaken at the University of Oxford on a Sercon 2022/Europa EA continuous flow isotope ratio mass spectrometer (CF-IRMS) system, together with alanine control standards, and multiple replicates of two internal standards (cow and seal bone collagen) to calibrate values to the relevant isotopic scales and to correct for scale-compression effects (Coplen *et al.* 2006; see Chapter 5 for full analytical details.

The results from the faunal bones are given in Table 5.4 and Figure 5.3 and are summarised in Figure 8.1. The results from the two faunal datasets show some significant differences, with the town site samples showing generally higher $\delta^{15}N$ values and more depleted $\delta^{13}C$ values. All samples show $\delta^{13}C$ values compatible with a C3 plant dominated ecosystem.

The values for the four hare samples from Tomb 159 show the typical low $\delta^{15}N$ values shown by many wild lagomorphs, while the sheep and goat are elevated in comparison, possibly suggesting that they were foddered on managed and manured pasture. Two of the sheep/goat samples from the town site show similar $\delta^{15}N$ and $\delta^{13}C$ values to those of the sheep from Tomb dromos, however the remaining sheep/goat samples from the town site show significantly different values, being elevated in $\delta^{15}N$ but depleted in $\delta^{13}C$, suggesting that the majority of sheep/goat samples from the town site come from animals that were grazed in a different location to those analysed from the tomb. The cattle from the town site show similar $\delta^{15}N$ and $\delta^{13}C$ values but with slightly more elevated $\delta^{15}N$ and slightly more depleted $\delta^{13}C$. Taken together, these data suggest that both cattle and the majority of the sheep from the town site were grazed in a more closed environment, where a canopy shading effect led to more depleted $\delta^{13}C$ values in fodder that was subsequently passed onto the grazing fauna from the town site. The two pig bone samples from the town site show very different values to the herbivore fauna with elevated $\delta^{15}N$ and $\delta^{13}C$ values, and they plot close to the range seen in the human bone collagen from the cemetery, suggesting that pigs may have been foddered on domestic waste rather than on agricultural or wild forage.

Minerals in the environs of Armenoi

Holley Martlew

Geologist Andrew Giże was to contribute a chapter to this volume but sadly died before it could be completed. The information below is taken from his chapters in the first volume on Armenoi (Giże 2018a; 2018b; Tzedakis *et al.* 2018) and from Martlew's field notebooks (1995–).

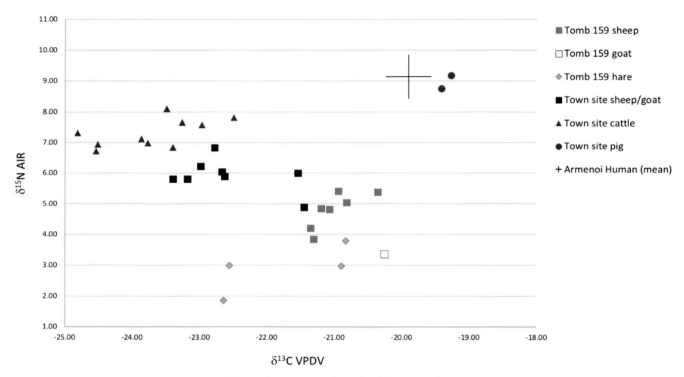

Figure 8.1. Armenoi faunal stable isotope data.

Iron and copper

Giże concluded that quality ores were sufficient to meet the needs of the Late Bronze Age population of Armenoi. A large deposit of iron ore occurs at Ano Valsomenero, ca. 4 km from the Necropolis. Analysis by Sherwood-Dickinson *et al.* (2018, 247) indicated that the iron was adequate for the manufacture of the iron artefacts contained in the tombs where shapeless lumps of iron have been found. Giże thought they represented early experiments in smelting on the part of Minoans. A rare copper source was located by Giże on the south coast but its potential as a source of ore has recently been superseded by the discovery of a copper mine, also at Ano Valsomenero (see Chapter 9).

Clays and decorative pigments

Giże identified good quality clays suitable for pottery manufacture on the south side of the village of Kastellos, that is, in the area of the 'city'. Yellow, orange and red to black pigments were used to decorate pottery and larnakes. Sherwood-Dickinson *et al.* (2018, 247) concluded that the colours were achieved by increasing the number of layers of iron oxide-hydroxide applied to the pottery before firing. Pockets of these are commonplace in the vicinity of the village of Kastellos and the Necropolis. In addition, a known source of haematite-goethite at Ano Valsomonero would

have been ample on its own to supply decorative pigments needed by the potters.

Quartz and soft stones

On the hillside on the north-east edge of the village of Kastellos, between it and the Necropolis, lies a Quartz Field. It consists of a large outcrop of calcite that has been replaced by quartz, as well as an abundance of Minoan sherds: the largest group found in the surveys that took place in 2001, 2002 and 2007. The crystals found included one of gem quality and a quartz tool made of two pieces of natural large crystals that had been water-rolled in a stream (Masters 2018, fig. 4.1; Martlew *et al.* 2018, 79, 99; Sherwood-Dickinson *et al.* 2018; see Chapter 9)

The Necropolis has yielded 160 seals, in roughly 80 (5%) of tombs. In most part they are made of soft stones found in the area and, with dates that are contemporary with the Necropolis, they are almost certainly the product of local craftsmen (see Becker 1976 for soft stone sources on Crete). They are made from a variety of stones: steatite, serpentine, soapstone, ophiolite and limestone. An outstanding example is the steatite heirloom pendant (R.M.S. 179-Azg) with a Linear A inscription, from Tomb 200. It dates at the latest to Late Minoan IIIB (Tzedakis and Kolivaki 2018, 15 and fig. 1.148). An outcrop of serpentine occurs on the road south to Spili where the greenish-black rock lies on a hillside in a huge bed and includes thin spikes and boulders large enough to carve objects like a stone vase. An outcrop of yellow-green ('chartreuse') soapstone occurs close to the serpentine 'hill'; ophiolite, a moderately hard stone, can be found near Spili, about 17 k south of the Necropolis. An example of its use is the stone axe found in the Bee Garden excavation (Trench 3) (Chapter 9, Fig. 9.18, below) and most limestone artefacts from the Necropolis were probably made from rock taken from an area to the south of the site. The necklaces studied appeared to be from a local source, most especially one which was of local sparite.

Semi precious stones: Jasper

To the south of the Necropolis is a vein of gem quality crystal with albite in many different shades. These can be clear or smoky. Giże identified several examples, including one from the Quartz Field.

List of minerals found in the area of Rethymnon and Chania

In addition to those already mentioned, during the course of his research Giże identified sources of the following minerals that had proven use within the 'city' and in the tomb deposits, or potential for identification in future: calcite, chamosite, 'chamosite-clinochlore series', 'chlorite group', chloritoid, clinochlore, epidote, gypsum, heamatite, magnesiocarpholite, muscovite, var. phengite, paragonite. Clearly there was a wealth of minerals in the area of the Necropolis and on the road south available to the inhabitants of Armenoi. This potential alone could have accounted for the location and wealth of the settlement here (Chappell and Allender 2018, 28), the 'city' of Armenoi, which we now believe to be referred to as *da-*22-to* in Linear B.

Andrew Giże on other subjects

Giże undertook a variety of additional studies made many important observations. For example, the Minoans' mining skills and and knowledge of rocks and minerals in the surrounding area, such as an understanding of how limestone fractures, could be seen reflected in, for instance, his observation that 70% of the steps and chamber entrances on the lower level of the Necropolis lie parallel to these natural fractures. This explains their orientation.

Giże was able to differentiate Armenoi workshop products of pottery from others on the basis of the clays employed. For example, the clay used for alabastron R.M. 20906/12 contained small black clasts (as well as one rare quartz clast). This fabric was typical of pottery produced in the Armenoi workshop(s). He suggested non-destructive testing of pigments on pottery and larnakes which could then be checked with pigment samples found in the Bee Garden excavation.

Necropolis and tomb construction

Giże made a number of important observations. A bed of conglomerate (a clastic sedimentary rock containing rounded fragments more than 2 mm in diameter) underlies the Necropolis. It is very irregular in shape and distribution and was avoided as far as possible by the builders who dug the tombs. However, it was incorporated into some structures, such as the wall of Tomb 221. As the ground runs east, the bed of conglomerate runs closer to the surface and was thus less easily avoided, which may be one reason why why the Necropolis is in two sections. Close examination allowed Giże to note the following:

- Tomb 213 is built just above the conglomerate which forms the floor of the tomb.
- Tomb 1 used the bottom of a fissure for the floor.
- Tomb 171 has a diagonal fissure at ca. 1 m height. If you put a hand in the hole you can feel the draft.
- Tomb 17 had a curved roof inside like a medieval chapel.
- Tombs could fail. Tomb 90 was misshapen. This was due to a small fault in the rock.
- Tomb 87 was complicated to build because of erosion.
- Tomb 206 has chisel marks on one side only and the wall is rough on the other side. It is not clear if this is represents natural weathering or was deliberately treated.
- The sizes of chisels used in construction varied. for instance, in Tomb 213 a small chisel, the size of a finger, was used.
- Giże identified vertical 'tidying up' chisel marks on dromos walls (*e.g.* Tomb 24).
- He called the wall at the entrance to Tomb 13, which was so refined, 'a labour of love' by an extremely competent craftsman. He commented that this man would have polished the stone if it had been possible.
- He thought that the well-built tombs at the top of the hill could have been built by the same skilled craftsman as he believed he could detect a 'hand'. Tombs 80, 79 and 175 could have been the same hand.
- The patterns of the chisel marks on the walls of Tomb 192 indicate the ease, or otherwise, of digging into the rock. Giże commented that the tombs on the lower slope look scruffy compared with those that run at the top, and that this was due to the type of rock. Tomb closures were easy at lower levels because one could just pick up rocks close-by. He felt the earliest tombs were the easiest to build (ie, softer stone).

Ultraviolet light study of seals

Giże examined a series of seals (see above) under ultraviolet light and comparing them with raw material from the nearby 'Quartz Field'. He noted a number of interesting variations.

- R.M. Σ73 and 78: the calcite is a perfect match for the Quartz Field.
- R.M. Σ49: the calcite could have come from the Quartz Field.
- R.M. Σ57, 62, 65, 66, 71, 79, 118, 119, 127, 133, 136, 169, 211: could also have come from the Quartz Field.
- R.M. Σ42: the calcite came from a source other than the Quartz Field.
- R.M. Σ97: the calcite was definitely not from the Quartz Field.
- R.M. Σ134: the seal is made of steatite.

Postscript: a tribute to Andrew Giże

This chapter has only been able to outline Giże's contribution to the project but the information is an important indicator of what he planned to present in detail concerning the mineral wealth that lay on the doorstep of the Minoans who founded the 'city' and built the Necropolis. The only thing I have left to report is his enthusiasm. Every time he took me on a field trip, his eyes were alight with pleasure as he talked about this part of Crete, which he loved and where he planned to retire. Everyone who has contributed to the study of the Late Minoan III Necropolis will join me in appreciation of the work carried out by, and the friendship of, a colleague we will never forget.

Bibliography

Becker, M.J. (1976) Soft-stone sources on Crete. *Journal of Field Archaeology* 3, 361–74.

Chappell, E. and Allender, S. (2018) Site investigations of the Necropolis and its environs: the search for the town. In Tzedakis *et al.* (eds), 23–52.

Coplen, T.B., Brand, W.A. Gehre, M. Gröning, M., Meijer, H.A.J., Toman, B. and Verkouteren, R.M. (2006) New guidelines for δ13C measurements. *Analytical Chemistry* 78, 2439–41.

Giże, A.P. (2018a) Geological setting. In Tzedakis *et al.* (eds), 213–30.

Giże, A.P. (2018b) Proposed method of tomb construction. In Tzedakis *et al.* (eds), 231–40.

Longin, R. (1971) New method of collagen extraction for radiocarbon dating. *Nature* 230, 241–2.

Martlew, H. Giże, A.P. and Kolivaki, V. (2018) Minoan diagnostic pottery from field and geophysical surveys: 1992, 1997, 2001, 2002 and 2007. In Tzedakis *et al.* (eds), 67–110.

Masters, P. (2018) Geophysical survey: Necropolis and town. In Tzedakis *et al.* (eds), 53–65.

Sherwood-Dickinson, C., Droop, G. and Giże, P.A. (2018) Ano Valsamonero iron deposit: a potential metal source for the Late Minoan III community In Tzedakis *et al.* (eds), 241–8.

Tzedakis, Y. and Kolivaki, V. (2018) Background and history of excavation. In Tzedakis *et al.* (eds), 1–18.

Tzedakis, Y., Martlew, H. and Arnott, R. (eds) (2018) *The Late Minoan III Necropolis of Armenoi* I. Philadelphia PA, INSTAP Academic Press.

The archaeological evidence which supports the identification of the Late Minoan III city of Armenoi as *da-*22-to*

Holley Martlew and Yannis Tzedakis

This chapter is presented as a response to Louis Godart's proposal (Chapter 7) that the 'city' of Armenoi is the one referred to in Linear B as *da-*22-to*.

1. The 'city' of Armenoi

Foundation

The foundation of the town/city that built the Necropolis was not happenstance. When the lowest levels of the excavations described below yielded pottery with the same chronology, Late Minoan IIIA:1, as the earliest finds in the Necropolis, it provided the tool that allowed us to identify the related settlement. It also became clear that the foundation of the town was the result of a deliberate decision, moreover that the founding fathers were shrewd and had moved with forethought, but why? What was the original impetus to look for a location and found a settlement, a city? There was one clear imperative above all. Because of the rise of the Mycenaeans, and the first Mycenaean invasion of Crete, trade south and the import of copper from Cyprus to Crete had ceased at the end of Late Minoan II, 1460–1450 BC. As a consequence, the flow of trade in Crete reversed from north–south to south–north on the main passage in western Crete. A town at the top of the escarpment at the north end of this passage, not far from and overlooking the sea, could take command of this new configuration – a vantage point not lost in more recent times, on the Germans, when they occupied Crete during World War II, who placed three anti-aircraft emplacements within the Necropolis (one damaged Tomb 55 as it was partly built over; see Fig. 6.1) and a munitions dump between the Necropolis and the village of Armenoi. The local commander took over a house at the eastern end of the village of Kastellos. Fortunately the Germans did not discover the Necropolis (see also Chappell and Allender 2018, 28–9). As Giże's work on the project has

indicated (see Chapter 8), it seems clear that the Minoans who settled in the area were also well aware of the finer points of choosing this location: local mineral wealth, local geology and favourable climate.

That was the motivation, but who were these founders? Such a decision, such an enterprise, would not have been undertaken by neophites but by sophisticated pioneers. The riches of the Necropolis, its very existence, indicate that those who built it intended to live in an independent entity that they themselves had created with the purpose in mind of making it great. We might suggest that the founders were men who saw an opportunity, a band with a leader, a (younger) son of the ruler of a polity in the west of Crete, Kydonia perhaps, with its long history. It could have been a group who lived on the south coast and realised that, unless a bold step was taken, their livelihoods were doomed. The existence of the Necropolis meant that the city that built it had to be nearby. Over the years small surveys to find a settlement had been carried out, but nothing definitive had been found. In 2003, a large team was assembled to carry out a major field survey which would cover an area within a 4 km radius of the Necropolis.[1] The survey, in various forms, continued until 2010.

Geological survey

Andrew Giże carried out a geological survey over many years. It extended much further than the immediate environs of the Necropolis east–west, north–south and along the south coast (see Chapter 8; Giże 2018a; 2018b; 2018c).

Geophysical survey 2001–2006 and subsequent excavation

Eileen Chappell and Steve Allender directed a field survey that was originally to be within the 4 km radius but was extended much further, from Atsipadhes to the Galliano

Gorge and from Vrysinas mountain to Valsamonero (Chappell and Allender 2018). Pottery from all periods was collected in the surveys (Ariotti 2018; Martlew *et al.* 2018). Late Minoan III sherds were found to predominate close to and in the modern village of Kastellos, which is located on a hill ca. 1 km above the Necropolis to the south-west, in an area that extended as far as the church of Aghios Fanourios.

In 2007, on the basis of an analysis of the results of the surveys, excavation started on the southern edge of Kastellos. In that year, under the auspices of The Holley Martlew Archaeological Foundation, the remains of the 'city' that built the Necropolis were finally located, in three areas on the outskirts of the village. The three areas were Armia (Tzanidakis Farm), the Bee Garden and the Quartz Field. The earliest finds in the excavations date to Late Minoan IIIA:1 – the same date as the earliest finds in the Necropolis. This indicates that the town became wealthy enough within a very few years to conceive and to start building what would become a splendid Necropolis.

The 'city' of Armenoi, however, is mostly buried beneath the modern village which was, in turn, built over a Venetian town and, as such, is under a preservation order. The unfortunate consequence of this is that much of the Minoan 'city', architecture and its artefacts, will never be known.

Armia, Tzanidakis' Farm

This area is located on the southern edge of Kastellos (Figs 9.1 and 9.2). A large farmhouse, constructed of limestone, was discovered. A road ran down the side of the building and a smaller one across the front. The entrance opened into a wide corridor with large rooms on either side. On the right side, facing, was a square room identified as a kitchen with the floor covered in gravel. On the left, a second room contained two large, partially sunken, decorated pithoi (Figs 9.3–5). Limestone tiles that had fallen from the roof were scattered across the room. Scanty remnants of a staircase led to what would have been the first floor. The finds were predominantly coarse-ware pottery dated to Late Minoan IIIA1–B2. Important finds include:

a) Clay sealing. Pit 1 (Fig. 9.6).
b) Kylix. Pit 1, Vase 1, 7 sherds. No paint remains. Max. D.: 15.6–8.0 cm. From the same chronological period as the founding of the Necropolis, dated to Late Minoan IIIA:1 (Fig. 9.7).
c) Mug. Pit 1, Vase 3, 7 sherds. Black paint remains inside/ outside. H.: 9.0 cm.; Base D.: 6.0 cm. Late Minoan IIIA:1, Armenoi workshop (Fig. 9.8).
d) Clay pounder. H.: 7.5 cm (Fig. 9.9).

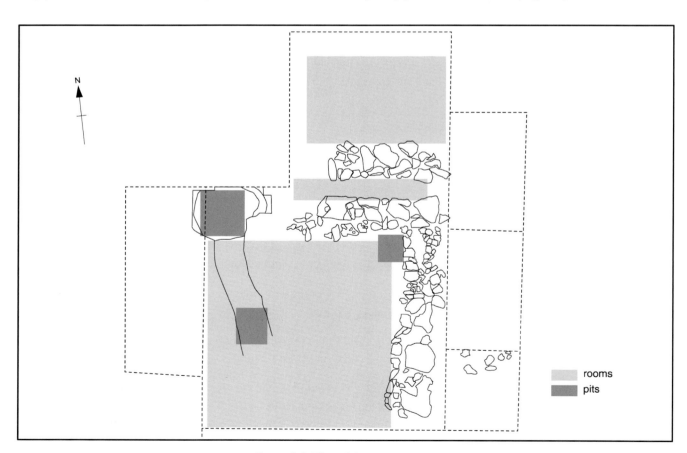

Figure 9.1. Plan of Armia excavation

e) Handle from a vessel used to serve food (Fig. 9.10).

f) Several diagnostic tripod cooking pot legs, all dated to Late Minoan III.

Figure 9.2. View of the Armia excavation from the north.

Figure 9.3. Armia excavation: decorated pithos 1.

Figure 9.4. Armia excavation: decorated pithos 2 emerges from the ground.

Minoan room and Minoan farmhouse(s)

In 1984–1985, a Minoan room was found outside Kastellos. Part of it had fallen away due to subsidence. Many sherds of pithoi were found in and around it. In view of the discovery of the farmhouse at Armia, the discovery of the remains

Figure 9.5. Armia excavation; decorated pithos 2.

Figure 9.6. Armia excavation: A clay sealing.

Figure 9.7. Armia excavation: Kylix. Pit 1 (Vase 1). From the same chronological period as the founding of the Necropolis: LM IIIA1.

Figure 9.8. Armia excavation: Mug. Pit 1 (Vase 3). Armenoi workshop. From the same chronological period as the founding of the Necropolis: LM IIIA1.

Figure 9.9 Armia excavation: pounder made of clay.

Figure 9.10 Armia excavation: handle from a vessel used to serve food.

of the room assumes greater importance as do the other possible 'farmhouses'.

In 1985, while excavating tombs, a worker told Tzedakis that he had found many sherds, mainly pieces of big pithoi, near the remains of walls near the cliff of Kastellos. He said that farther away, on the fertile plateau approximately 150 m below Armia, he had been told there had been a Minoan farmhouse that had been completely destroyed by a farmer. This means that in the fertile area outside the town there must have been at least three farmhouses in the area belonging to members of the Minoan community, Armia and two others. The Minoan room could indicate there were four.

Bee Garden: area on the north edge of Kastellos

The distance to the Necropolis is approximately 1 km and, between Armia and the Bee Garden to the north, is slightly more than 1 km (Fig. 9.11). Three trenches were opened and a quantity of Late Minoan III pottery, specifically Late Minoan IIIA1–IIIB2, and pithoi sherds were found:

a) Pithos sherd. Relief decoration outside. Max. D.: 5–8 cm.; Th.: 2 cm (Fig. 9.12).
b) Pithos sherd. Relief decoration outside. Max. D.: 4–10 cm.; Th.: 1.2 cm (Fig. 9.13).
c) Pithos sherd. Curved relief decoration outside. Max. D.: 6–7 cm.; Th.: 2.0 cm (Fig. 9.14).
d) Small pieces of obsidian were found between Trenches 1 and 3.
e) Charcoal was found in Trench 2.

In Trench 2, a stone shelf was discovered which appeared to have been used for butchering; animal bones were found on top. Twenty faunal bones were submitted for analysis (see Chapter 8). Additional excavation revealed a large house made of white and pinkish limestone. It had been built on two levels connected by a stone staircase (Fig. 9.15). Tzedakis recognised that it was a typical Minoan staircase with steps of depth 17 cm. On the north side, on the lower level there was a long corridor in the centre of which were found a number of conical cups. To the left of the corridor there was a room with a rich scatter of pigments. It was unquestionably a pigment workshop (Fig. 9.16).

Giże located the likely source of these pigments along a path that led around the hillside between the Bee Garden workshop and the Necropolis. Pockets of iron oxides-hydrocides occur in limestone but it is the only time, to our knowledge, that a pigment workshop has been found near such a deposit (see Chapter 8).

Other finds included:

a) Stone axe. Ophiolite. It was not made of a local stone, according to Giże. The nearest source of ophiolites would be near Spili, ca. 17 km from the excavation (Fig. 9.17; see Chapter 8).
b) Wall sherd with a hole which would have been used for a rope handle (Fig. 9.18).

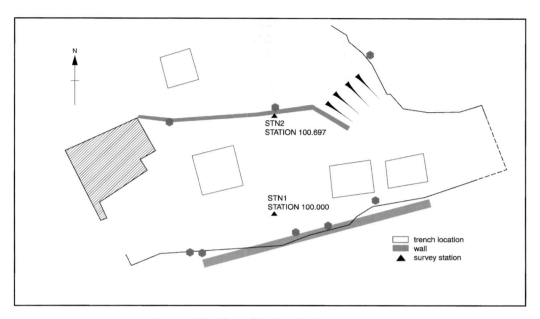

Figure 9.11. Plan of the Bee Garden excavation

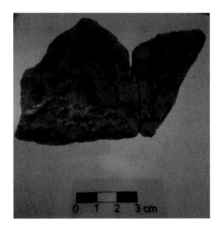

Figure 9.12. Bee Garden excavation: sherd with impressions.

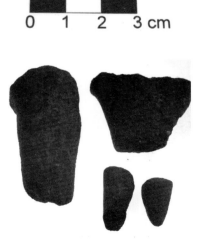

Figure 9.13. Diagnostic tripod cooking pot legs, Late Minoan III.

Figure 9.14. Bee Garden excavation: pithos sherd with curved relief decoration.

c) Sherd with incised lines (Fig. 9.19).
d) Metal skewer (Fig. 9.20).
e) Diagnostic tripod cooking pot legs were found, all dated to Late Minoan III (Fig. 9.13).

The upper terrace of the building had substantial footings of wall foundations which abutted the cliff edge of the terrace. The geophysical survey (see below) reported a quantity of pottery just below the surface and further excavation was strongly recommended. It became clear from the finds that the building housed Minoan workshops and that outside it was a butchery station. The period of use was the same as the Necropolis.

Quartz Field

This lies on the north side of Kastellos, on a hillside that leads down to the Necropolis. Giże located an abundance

Figure 9.15. Bee Garden excavation: typical Minoan staircase.

Figure 9.18. Bee Garden excavation: wall sherd with a hole which would have been used for a rope handle.

Figure 9.16. Bee Garden excavation: the scatter of pigments.

Figure 9.19. Bee Garden excavation: sherd with incised lines.

Figure 9.17. Bee Garden excavation: stone axe.

Figure 9.20. Bee Garden excavation: metal skewer.

of Late Minoan III sherd material and a large outcrop of calcite which was being replaced by quartz (Giże 2018b; Chapter 8). The potsherds found were numerous, one of the largest groups found in the Surveys (Martlew *et al.* 2018) but, in terms of defining the mineral wealth of the 'city' of Armenoi, the most significant finds on the hillside were the quartz and calcite which provide further testimony to the mineral wealth in the immediate environs (Giże 2018b; Sherwood-Dickinson *et al.* 2018, 243–8).

The lowest level in all the excavated areas was dated to Late Minoan IIIA:1, the same as the earliest finds in the Necropolis. The Minoan 'city' clearly met with success from the outset and almost immediately it was decided to spend some of its new-found wealth in honouring its citizens by the building of a majestic Necropolis, maintaining and expanding it until both were eventually abandoned (Tzedakis and Kolivaki 2016).

Geophysical Survey 2010

The survey was carried out by Peter Masters after three areas of the 'city' had been identified. Earth-resistance survey readings were undertaken with a twin-probe configuration using a Geoscan RM15 instrument and readings were logged at 1.0 × 0.5 m intervals (Waker 2000). A gradiometer survey was also carried out using a Bartington Grad 601-2 fluxgate gradiometer (Bartington and Chapman 2004). The terrain made the survey difficult (Masters 2018, esp. p. 54 and fig. 4.1).

Areas near the Necropolis

Area 4, on the east side of the Necropolis, revealed the remains of a kiln and ferrous iron or concrete. A paved road was also identified which could have led into the Necropolis. Area 5, to the east of Tomb 213, showed features that might indicate the presence of tombs. Pits or ditches were identified. Three areas were examined that could have revealed more tombs – two had positive results, east of Tombs 73 and 74 and north of Tomb 166.

The Bee Garden was recommended for further excavation as was Armia. The Pigment Field (to include Quartz field), Area 11, may have part of a wall foundation. Ground-Penetrating Radar (GPR) was highly recommended.

2. The Late Minoan III Necropolis of Armenoi (Late Minoan IIIA:1–Late Minoan IIIB:2 ca. 1390–1190 BC)

Built of white limestone on a sloping hillside, the Necropolis would have been highly visible to the many who travelled the only north–south passage in western Crete, in the years it was in use, and it would have remained so, for years thereafter. This is almost certainly the reason that at the end of the Minoan period when the Necropolis was abandoned, a few were plundered, one of which was the royal tomb,

Tomb 159 (Fig. 9.34). The most important evidence for the identification of the 'city' of Armenoi as *da-*22-to* in the Linear B tablets is therefore the existence the Necropolis, unique and extraordinary in the fact that it survived as a complete and almost entirely intact cemetery until its rediscovery in 1969.

This discussion will start with a close examination of the Necropolis. It will bring into focus why it had to have been built by an important and wealthy community: inhabitants of a city.

Modern name and more recent history

Armenoi is not an ancient name but was taken from the nearby modern village of Armenoi. There are three villages with this name in Crete, that near the Necropolis, one between Rethymnon and Chania and one in eastern Crete. The names were given to settlements established in the 10th century by the general, later emperor, Nikophoros II Phokos, reigned AD 963–969). He liberated Crete from the Saracens in 961. A ferocious soldier known as the White Death of the Saracens, he had come from Cappadocia, part of what was then Armenia, much larger than modern Armenia. The general resettled people in Turkey and Crete, not simply to give them the opportunity of building better lives but to build up the Byzantine Orthodox presence in an island which had been conquered by the Muslims in the 820s.

Winning the Battle of Chandax in AD 961, the main Muslim fortress and capital of the island, was a major achievement by the Byzantines. There had been many failed attempts. The victory restored Byzantine supremacy over the Aegean and far diminished the threat of Saracen pirates who had been using the island as a base for their operations in the eastern Mediterranean.

The date of foundation of the Necropolis: Late Minoan IIIA:1 ca. 1390–1370 BC

It is important to note that the founding of the 'city' and Necropolis dates to a time before the fall of Knossos in Late Minoan IIIA:2, ca. 1370–1340 BC. Imported vases found in various tombs indicate that there was an extensive trade relationship between the two sites. One of these is Tomb 89, in which skeleton 89B, a person who was born and lived a great part of his life elsewhere than Crete, was buried. The finds included vases which were imports from the Knossos workshop (see Chapters 4 and 6).

The nature of relationship between Knossos and the 'city' that built the Necropolis, during and after the fall, during which time life at the 'city' of Armenoi appears to have continued as normal, raises many intriguing questions but here we focus on the wealth of the Necropolis and the reason why it had to have been built by inhabitants of a wealthy 'city'. In instances where the information mirrors or amplifies Godart's analysis, reference is made to the relevant Linear B tablet.

Figure 9.21. The Late Minoan III Necropolis of Armenoi viewed from the northwest.

Size: what has been excavated extends over a large area, is ca. 0.11 ha (1100 m²).

Type: The majority of tombs are built graves, chamber tombs with corridors/dromoi, stairs or ramps (this is determined by geology), that lead to chambers.

The Late Minoan III Necropolis of Armenoi represents a cultural mix: the existence of chamber tombs indicates a Mycenaean influence; whereas the use of larnakes (sarcophagi) to bury the dead was Minoan in origin.

Chamber tombs: To date 232 built structures have been excavated. Most (179 = 77.15%) are chamber tombs, the remainder are without chambers. The 179 range from tombs with small chambers to those which have large and elaborate chambers. Examples of the latter are Tomb 24 with a pillar and Tomb 159 with a bench and a

pilaster (Tzedakis and Kolivaki 2018, figs 1.15 and 1.16, respectively).

It is clear, therefore, that it was a hierarchal society but equally clear that the rulers of Armenoi valued all their citizens, as exemplified by the scientific results discussed in Chapter 4 and 5 which indicate that all the citizens shared the same diet, high in terrestrial proteins and with no quantifiable consumption of fish.

Human skeletal remains: it is estimated that, originally, the chamber tombs would have held the remains of approximately 1000 people. Human skeletal material survived in 50.83% of the chamber tombs. This does not mean the rest never had any burials inside them. The construction of the Necropolis in limestone, the climate, taphonomic processes, all led to the disintegration and disappearance of skeletal material over time. It is why there are no complete skeletons.

What did the Minoans of the 'city' of Armenoi look like?

A photograph of a skull that was reconstructed for the International Exhibition, *Minoans and Mycenaeans Flavours of their Time*, is reproduced to give an indication of what one individual may have looked like (Tzedakis and Martlew 2003, 243, fig. 212; Fig. 9.22). It is that of a male found in Tomb 132. He had been buried with a woman in the only larnax known to have held more than one person. The larnax, R.M. 22608, was decorated with horns of consecration.

What did they wear?

Information on clothing comes mainly from decoration on larnaxes buried in the tombs. On Larnax R.M. 1707 from Tomb 139, dated to Late Minoan IIIA:2 (see back cover), hunters are shown wearing short tunics while participating in a 'sacred hunt'. The fabric (woven?) has criss-crosses and vertical diagonals on two men and horizontal ones on a third. A hunter on larnax R.M. 1712, from Tomb 24, wears a zoma, a Minoan loin cloth. A goddess (or priestess) wears a long robe covered in elaborate undulating concentric triangles. She appears to have sleeves that hang down in long out curving spirals, as illustrated on larnax R.M. 1706 from Tomb 24, dated to Late Minoan III B:1 (see back cover). What is clear from the larnakes, as in the frescoes, Minoans did not wear simple plain fabrics whatever their occupation (hunter, celebrant, or goddess). They dressed for the occasion.

The hunters have short hair which is in disarray. One hunter is brandishing a double axe. They appear to be wearing a moccasin type of shoe. The 'goddess' is wearing a long flowing ponytail. Her hands are raised in supplication.

Figurine (bust), ceramic

R.M. 9244, Tomb 159, the royal tomb, LM IIIA/B: the man wears a puffy hat which rises to a stiff point in the centre (see Fig. 9.33a, below). It has a small protrusion/beak at the front. The top of the hat is decorated. Lines radiate out from the large circle at the centre. Does this represent the sun? It is not a crown as we understand it. It is more like the *corno ducale* of a Venetian doge which, in turn, is reminiscent of a Phrygian cap or, more importantly considering the time period, the white crown of Upper Egypt. One even might ask, considering where it was found, is this the headgear of a king? (Fig. 9.33b). Whoever the figurine represents, king or layman, if the decoration on the neck does represent gold chokers, it is one of only a few clues (considering the wealth of the Necropolis) to the style of (gold) jewellery worn at the site and, as the tomb was looted, could mirror what was once within it (see below under Tomb 159 for further discussion and Chapter 10 Postscript). There is one splendid example of gold jewellery found at the Necropolis, a delicately carved gold leaf pendant, also found in Tomb 159 (R.M.M. 3241; see below and Appendix).

Figure 9.22. Reconstruction of male skull from Tomb 132 (after Tzedakis and Martlew 2003, fig. 243).

Enigmatic structures

There are built structures without chambers in the Necropolis. They were given tomb numbers. Those in **bold** have niches: Tombs 7, 9, 12, **15**, 21, **28**, **30**, 33, **50**, **51**, 53, **57**, 66, 68, 70, **72**, 81, **96**, **97**, **105**, **106**, **112**, **113**, **116**, 122, **128**, **131**, **134**, **135**, 138, **141**, 142, **147**, **151**, **152**, **153**, **166**, **176**, **182**, 185, **195**, **219**, **222**, 223, **224**, 228, 231, 232. They number 45 out of a total of 232 (ca. 20% of the tombs) and 32 of the 45 (ca. 71%) have niches. Their function as tombs is not certain. Giże suggested that the configuration and condition of the geology of the area could be part of the reason for tombs not having chambers. Only seven of them (Tombs **51**, 68, 70, 122, **134**, 136, 142) have produced scrappy remains of bones, not enough skeletal material to analyse fully other than to say they were not child burials. The problem of built structures without chambers is currently under further investigation (Martlew forthcoming).

Wealthy tombs

Tombs 8, 10*, 13, 20, 24*, 45, 52, 55, 60, 67*(small tomb but very rich), 115, 118, 140*, 159* (largest, richest tomb), 165*, 167*, 179, 184, 186, 198, 203, 206* 211 and 213* are considered to be wealthy. These 24 account for ca. 10% of the 232 chamber tombs. The nine tombs (37.5% of wealthy tombs) marked with an asterisk (*) were especially rich and represent 3% total. The largest and richest, Tomb 159, is described below. These may be considered to contain elite burials but this does not mean however, that the others were poor: the great number of excellent vases and bronzes that were found throughout the Necropolis attests to this. The lack of overlap between 'rich' tombs and those from which samples of human skeletal material were taken is because the researchers did not choose material on the basis of the type of tomb but chose the best-preserved teeth and petrous bones.

Religion and ritual

It is clear that the people of Armenoi practised a religion which believed in burial rituals and performed them both inside the chambers, as evidenced by the types of vessels found in the tombs, and outside in the dromoi/corridors that led to the chambers (again evidenced by the types of vessels found). The types of vessels inside the tomb differed from those found in the dromoi but both groups provide evidence for rituals. Incense burners, alabastra, stirrup jars, jugs and miniature vases occur inside tombs; kylikes/goblets and coarse sherds of jars and cooking pots predominate in the dromoi (Martlew and Kolivaki in prep.).

A number of pits were identified as ritual pits on the basis of the types of vessels which would have been used in rituals. They included kylikes/goblets and cooking vessels. Examples include a group of pits to the south-west of Tombs 177, 178 and 181 (Tzedakis and Kolivaki 2018, fig. 1.8); a pit north of Tomb 225 and one east of Tomb 202.

Skeletal pathology, stable isotopes and ancient DNA

As Darlene Weston's work has shown (Chapter 3), the people from Armenoi suffered from common ailments, by far the most frequent being osteoarthritis of the spine, dental disease, fractures (most frequently of the wrist), anaemia and non-specific infection. There were a few less common pathologies: a congenital hip dislocation, a case of sinus infection, and one possible case of tuberculosis. There was also evidence that a number of the Armenoi individuals were engaged in physical labour that caused enlargement of the muscles. There are no clear indications of differential status among these results. One familial association was recognised from an unusual feature in the jaws of two individuals from Tomb 159.

The small sample of individuals subjected to isotope and ancient DNA data (Chapter 4 and 5) has identified a number of outliers – that is, people who were either not born on Crete and/or spent some of their early lives elsewhere – indicating the cosmopolitan composition of the Armenoi population, and at least four familial relationships including the presence of three generations of one family within Tomb 203. These results are highly significant for developing our understanding of the population of Crete at this period.

3. Contents of the tombs

Larnakes/sarcophagi

The production of imposing, large, decorated ceramic chests could only be undertaken by a prosperous town. The elaborate painted designs, all of which relate to ritual, indicate they were specially commissioned. Thirty-three larnakes/sarcophagi were found. Ten are still to be restored. All are decorated. Two larnakes painted in polychrome were found, in Tombs 10 and 24. The larnax from Tomb 24, (R.M. 1712; see back cover) depicts a hunter standing on the back of an agrimi (Cretan wild goat), a dog and a net (Tzedakis and Kolivaki 2018, fig. 1.17). That from Tomb 10, (R.M. 1709, see back cover) depicts motifs that include partridges, deer, fish, Minoan flowers, double axes and horns of consecration (Tzedakis and Kolivaki 2018, Fig. 1.18). A third polychrome larnax was found in West Crete at Dramia (Kanta 1980, 237). Two of the three polychrome larnakes extant were found in the Necropolis.

Tomb 159, the royal tomb, was thought to have held two larnakes until July 2023, when Tzedakis identified a third (see Appendix). Two larnakes were found in Tomb 24 and in Tomb 55. The remainder of the tombs with larnakes each contained only one. Tzedakis identified larnakes that potentially came from the same workshop as they appeared to have been painted by the same hand (referred to as the 'Master of Papyrus'; Tzedakis and Kolivaki 2018, 10 and fig. 1.25).

Pottery

In excess of 800 beautifully decorated vessels were found at the Necropolis. Most are intact and in pristine condition (Tzedakis and Kolivaki 2018, 13–15). Where pots were broken this was, in some cases, due to deterioration resulting from climatic conditions but mostly to the collapse of the tomb roofs. Most of the vases are shapes which would have been used in ritual (see Tzedakis and Kolivaki 2018 for a discussion of the pottery).

Bronzes

More than 300 bronze objects were found. These include a bowl, a dipper, daggers, knives, cleavers, spearheads, and razors (see Tzedakis and Kolivaki 2018 for a discussion of the bronzes). Finds included:

a) R.M.M. 386. Bowl. H.: 5 cm; D. rim: 12.4 cm. Tomb 35. LM IIIA (Fig. 9.23).
b) R.M.M. 593. Tweezers. L.: 5.5 cm; Th.: 1.7 cm. Tomb 74. LM III (Fig. 9.24).
c) R.M.M. 571. Dipper. H.: 12.4 cm. Tomb 115. (Fig. 9.25).

Seals

(Olga Krzyszkowska)
The number of seals found at the Necropolis provides clear testimony to the wealth of the 'city' which built it. More than 160 seals have been recovered from the tombs, a greater number than from any other Bronze Age Aegean cemetery (Krzyszkowska 2005, 212–214, nos. 417–420, 422–425; 2019, 492–493, pl. CLXXVIIb). Roughly 80 tombs, or about one-third of the total, yielded seals; tombs containing several seals (2–6) are common. Locally available soft stones, such as serpentine and calcite, were used for the majority of seals (see Chapter 8); these are mostly datable on stylistic grounds to Late Minoan IIIA1–2. In contrast, only three or four hard stone seals can be dated

Figure 9.23. Bowl (R.M.M. 386). Tomb 35. LM IIIA.

to Late Minoan IIIA1. These include an exceptionally fine lentoid of rare lapis lacedaimonius (R.M. Σ14 from Tomb 17; Tzedakis and Kolivaki 2018, 15–16, fig. 1.47; incorrectly dated to Late Minoan II). There are only about 50 such in existence, most of which are, or are likely to be, Cretan products, even though the stone was imported from the Greek mainland (the only source is near the village of Krokees in southern Laconia and indeed the ancient author Pausanias mentions a quarry there).

Of especial interest is the concentration of antique seals which had been produced hundreds of years prior to their deposition. These amount to about 20% of the total. Normally made of attractive hard semi-precious stones, these were perhaps valued chiefly as items of personal adornment and may have passed through many hands before reaching their final resting place. In addition, a few seals, such as examples of the Mainland Popular Group and Mitannian Common Style, point to overseas connections. Examples of seals are illustrated in Tzedakis and Kolivaki 2018, figs 1.45–1.48. Tomb 203, one of the wealthiest tombs containing individuals representing three generations of one family (Chapter 5 above), included one seal (see below).

Funerary stele (tombstones)

Eleven funerary stele were found, but none *in situ* (Tzedakis and Kolivaki 2018, figs 1.10–1.11).

Unique and important finds

Among the objects found in the tombs are a number that, except for the boar's tusk helmet (one of two), are so far unique in the Minoan world. These include:

a) R.M. Σ14. Tomb 17, ?inside larnax. Seal. Lentoid. Chlorite or serpentine. Heavily abraded; edges battered. D.: 1.72–1.75 cm. Late Minoan.

b) R.M.S. 194 Tomb 108. A carnelian bead which imitates a scarab. L.: 1.2 cm. LM IIIA.

c) R.M.Y. 583. Tomb 140. Pendant in steatite. Male figure with hands crossed on chest. D.: 1.0–2.5 cm. LM III.

d) R.M.L. 2929. Tomb 181, niche. Pendant in calcite, animal head with staring eyes. Max. D.: 1.1–1.08 cm. LM IIIA.

Figure 9.24. Tweezers (R.M.M. 593). Tomb 74. LM III.

Figure 9.25. Dipper (R.M.M. 571). Tomb 115. LM IIIA/B.

e) R.M.M. 3241. Tomb 159 (see Appendix) Leaf-shaped gold pendant. L.: 2.0cm. Dated to Late Minoan IIIA.

f) R.M. S179-Azg. Tomb 200. A steatite pendant with a Linear A sign. H.: 1.2 cm. It is an heirloom dated LM IB (Tzedakis and Kolivaki 2018, fig.1.48).

g) R.M.O. 266. Tomb 167. Boar's tusk helmet. H.: 19.6 cm (Tzedakis and Kolivaki 2018, fig. 1.50). LM IIIA:1–IIIB (Fig. 9.26).

h) R.M. Δ1. Tomb 187. Reed basket decorated with bronze nails. H.: 24 cm. Late Minoan IIIA:2/IIIB (Tzedakis and Kolivaki 2018, fig. 1.49).

Figure 9.26. Boar's tusk helmet (R.M.O. 266). Tomb 167. LM IIIA:1–IIIB.

Figure 9.27. Pithos sherd (R.M. 29818). Tomb 231. LM IIIA.

i) R.M. 29818. Tomb 231. Pithos sherd. Incised pattern on inner surface which had been executed with a tool. The exact pattern is not known. D.: 7.0–11.0 cm. It could be a game or even a plan of the town or the Necropolis. LM IIIA (Fig. 9.27).

j) R.M. 9422 Tomb 159. Human figurine (bust) in clay, with full hat (see Appendix and Chapter 10 Postscript (Fig. 9.33 below).

Tin-plated kylikes/goblets and a cup

These were found in several tombs. Excellent examples were found in Tomb 159 (R.M. 3455, 3459, 3460 (Fig. 9.28; see Appendix). The vessels were tin-plated to imitate silver and were imitations of imported Mycenaean tin-plated vessels made in both the Knossos and Armenoi workshops, some of which appear to have been made *after* the palace is presumed to have fallen. It is extraordinary that there could have been entrepreneurs at Knossos who continued to make pottery and to trade after such a disaster. If so, it would reveal a lot about the character of the Minoans. In spite of having their pre-eminent palace reduced to a ruin, the upheaval that had to have resulted, and repercussions at other sites such as Armenoi, the potters appear to have persevered with production.

Tomb 159 and Tomb 146

By far the largest and most impressive of the rich tombs, Tomb 159, illustrates clearly why the Necropolis of Armenoi is exceptional, as does Tomb 146, whose contents included the stirrup jar painted with the Linear B inscription, *wi-na-jo*. These two tombs underpin Godart's interpretation.

Figure 9.28. Tin-plated goblet/kylix (R.M. 3460). Tomb 159. LM IIIA:2.

Tomb 159: a royal family tomb

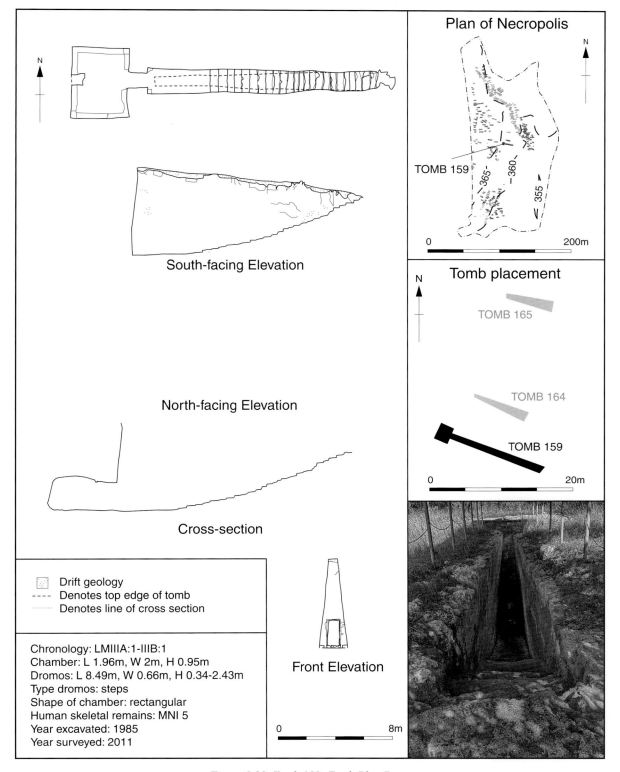

Figure 9.29. Tomb 159: Tomb Plan Page.

R.M. 3463

Figure 9.30. Base of an incense burner, large and coarse as one found in the dromos (R.M. 3463). Tomb 159. LM IIIB.

Figure 9.32. The coarse base of a large incense burner (R.M. 17272). Tomb 159, dromos. LM IIIB

R.M. 3457

Figure 9.31. Cup, tin-plated (R.M. 3457). Tomb 159. LM IIIA:2–LM IIIB:1.

Figure 9.33. a. Male protome. The eyes were added on separately. The left eye socket is empty, the right eye is complete with a hole for the pupil. There is a centre depression in the front where something had been broken off; b. hat on the male protome seen from above (R.M. 9422). Tomb 159. LM IIIA:2.

Godart cited the importance of a large and imposing princely tomb at the Necropolis as evidence that supports the identification of *da-*22-to* as the 'city' that built it. Tomb 159 (Fig. 9.29) gives every indication of being a regal tomb, one fit for a king. Tomb 159 is huge. It has a monumental staircase of 25 steps. The dromos/corridor measures 8.45 m in length. It had been robbed but the chamber still contained the remains of five people (four male and a foetus). Osteological analysis (see Chapter 3 above) revealed that at least two of the four males of those buried in Tomb 159 were related. A genetic relationship between skeletons 159Γ and 159Δ was identified due to the fact that they shared an inheritable severe overbite. They were either brothers or, more probably, father and son.

The tomb also contained three decorated larnakes and many important items including a tin-plated kylix (R.M. 3460; Fig. 9.28); The base of a large, coarseware, Late Minoan IIIB incense burner (R.M. 3463; Fig. 9.30); a tin-plated handled cup (R.M. 3457; Fig. 9.31); the coarse base of an incense burner, found in the dromos (R.M. 17272; Fig. 9.32) and the human male figurine (protome; R.M. 9422; Fig. 9.33). It dated to Late Minoan IIIA:2. The style of the figurine is distinctive. The eyes are open wide and outlined with a painted band. The object has suffered over time with the result that the left eye socket is empty. The right eye is complete with a hole for the pupil.

There is a centre depression in the front where the nose had been broken off. The eyes and nose are applied separately, in the same style as the protome on a double vase from the chamber tomb at Kalami (on the main road between Rethymnon and Chania). The figurine is 7 cm tall and 3 cm in diameter. It is wearing a full hat and around the elongated neck there is a wide zig zag (gold?) 'choker', and possibly plain, wide (gold?) 'chokers' above and below it and as discussed above, it might just represent the king himself. The figurine is unlikely to have stood alone because it has a tear at the base. Whoever it represents, considering its appearance and context, it must have had a ceremonial purpose and be part of the paraphernalia of a ruler (see Chapter 10 postscript for an interpretation).

The dromos of Tomb 159 is illustrated in Figure 9.34 where the two visitors give a good indication of the huge scale of Tomb 159 which lies to the left of the photograph, with Tomb 164 to the right (Fig. 9.35) illustrating how the sizes of the two tombs differ. The size and execution of Tomb 164 (see Fig. 9.36) compares well with the other important tombs in the Necropolis; they are all small in comparison with Tomb 159.

Figure 9.34. The dromos of Tomb 159 with two visitors from Austria standing inside (to provide scale).

Figure 9.35 Tomb 159 and Tomb 164.

A ritual feast

The quantity of sherd material found in the dromos was enormous, more than was found in the dromos of any other tomb. The size of the vases, most of them decorated, was also unique within the Necropolis (see Appendix; Martlew and Kolivaki in prep.). The collection of vases found in the dromos is impressive. Rituals had clearly taken place. Kylikes/goblets, jugs, and stirrup jars indicate drinking, most probably wine, even possibly the special cocktail, the mix of resinated wine, barley beer, and honey mead found in the sherd from the ceremonial pit near Tomb 178 (for Organic Residue Analysis, see Chapter 2). The base of the incense burner indicated that incense was used as part of ritual. Sherds of kraters and a pithos were found. The result on the 12 faunal bones submitted for analysis indicated the presence of sheep, a goat and four hares (with no pigs or cattle as recorded at the town site); it is an unusual and unexpected combination. More compelling is that the two faunal datasets, dromos and town site, showed significant differences which suggested that the sheep and goat in the dromos were raised specifically for ritual purposes (see Chapters 6 and 8). Our ambition is to submit sherds from the dromos for Organic Residue Analysis. In the meantime, to identify a ritual feast with its main components, that took place in the dromos of the royal tomb, is a great accomplishment, the result, as is this volume, of a partnership between archaeology and science.

Figure 9.36. Close up of Tomb 164.

Tomb 146

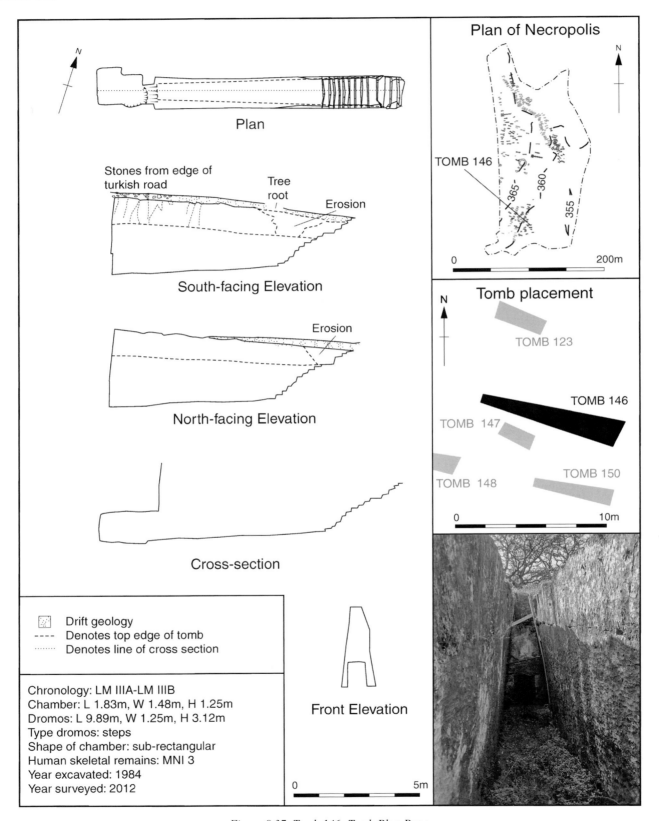

Figure 9.37. Tomb 146: Tomb Plan Page.

Godart cites (Chapter 7) the stirrup jar found in Tomb 146 which has a Linear B inscription. Tomb 146 (Fig. 9.37) is one of the largest tombs. The stirrup jar (R.M. 3363), painted with the Linear B inscription *wi-na-jo*, is one of the most important discoveries made in the Necropolis (Fig. 9.38). Two other stirrup jars with the same Linear B inscription, in the same hand, were found, at the Unexplored Mansion at Knossos H.M. 18374), dated to Late Minoan IIIB:1 (Fig. 9.39), and a house at Midea in the Argolid (Mi 95 AA1) dated to Late Helladic IIIB:2 (Fig. 9.40; Tzedakis and Kolivaki 2018, fig. 1.34). Stirrup jars found at sites on the mainland at Eleusis, Mycenae, Tiryns and Midea are known to have been made from clay from West Crete (Catling *et al.* 1980, 48–113).

The tomb contained a male(?) skeleton, 146A who may have been an outlier, a discovery worth noting in a tomb which contained a vase with a Linear B inscription (see Chapter 6 for discussion of skeleton 146 A).

The chamber contained the remains of one painted larnax (R.M.17209) of Late Minoan III date. The dromos contained a huge quantity of pottery sherds including the remains of three cooking pots, five stirrup jars, including R.M. 17211 (Fig. 9.41) and 13 kylikes/goblets (see Appendix).

Figure 9.40 Close-up of Linear B inscription. Midea, house excavation (Mi 95 AA1). LM IIIB:2.

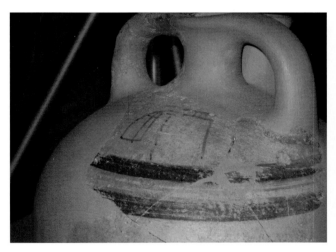

Figure 9.38. Linear B inscription on stirrup jar (R.M. 3363). Tomb 146. LM IIIB.

Figure 9.39. Stirrup jar with Linear B inscription from the Unexplored Mansion at Knossos (H.M. 18374). LM IIIB:1.

R.M. 17211

0   5 cm

Figure 9.41. Stirrup jar (R.M. 17211). Tomb 146, dromos. LM IIIA:2.

4. Correspondence with Linear B tablets cited as evidence by Godart

Godart discusses the **Co** series table of scribe 107 (**C 901, 989, 5544, 5733**) which mentions bulls and herds of cows (see below concerning cattle breeding). Bulls and/or their attributes are identified on larnakes R.M. 1710 and R.M. 5121 (see back cover). Another tablet (**Ga 1058**) mentions a religious procession (of bulls) and refers to a nearby peak sanctuary. Bull processions as part of (funerary) ritual are depicted on larnax R.M. 1710. In the Middle Minoan period peak sanctuaries were a feature of Minoan religion. By the Late Minoan period cult centres on mountain tops had gone of use. After their demise, however, they retained a 'sanctity' for the population. Two peak sanctuaries can be viewed from the 'city' of Armenoi, at Atsipadhes, located near Spili, ca. 66 km from the Necropolis at an elevation of 984 m (Peatfield 1994), and on Mount Vrysinas, which looks down on the Necropolis at a distance of about 1 km. Hundreds of small idols of bulls and human figurines have been found on its slopes. The existence of the latter has been interpreted as relating to health. Late Minoan III sherds have been found at the top as well as lower down, which show it was important to the Minoans of the time. Because of these finds it has been proposed that, in Late Minoan III when the 'city' of Armenoi had come into existence, Mount Vrysinas was used as a look-out post. This would have made sense because the mountain sits above the main north–south trade route that runs between it and the Necropolis and connects the north coast (Rethymnon) to the Libyan Sea (Aghia Galini; see Giże 2018a, 19–21). Figure 2.1 shows the location of Mount Vrysinas. Recently, in the spring of 2023, a third, a new peak sanctuary, has been found at Valsamonero, ca. 5 km from the Necropolis. The three peak sanctuaries form a triangle. From the top of Mount Vrysinas one can see the two others (see below). Valsamonero is βαλσαμο in Cretan dialect: the name refers to the balsam fir, the source of resinous products. As a paste it is used to treat cuts, burns and snow blindness and, as a tea, to induce sleep, relieve coughs, sore throats and sinus congestion. The tree can also be used for light frame construction.

Further evidence cited by Godart

A stirrup jar with the painted Linear B inscription *da-*22 -to*, found at Eleusis on the Mainland (**EL Z 1–2**, scribe 518), was an import from Crete. **EL Z 2** *da-pu2-ra-zo, wa*: the *wa*-ending is significant. Godart explains it is an abbreviation of an adjective which means 'belonging to the king', i.e. produced and exported under palatial authority (Chapter 7). The date is important here. The king cannot be at Knossos. The stirrup jar dates to Late Minoan IIIB:1, which is after the fall of Knossos which occurred in Late Minoan IIIA:2, hence the king/wanax to which the inscription refers has to be in West Crete. Kydonia/Chania became an important palatial centre after the fall of Knossos, but Godart emphasises that there were others. The outstanding finds at the Necropolis, as well as the architecture of the chamber tombs themselves (especially the royal tomb), gives every indication that the 'city' of Armenoi was one of these centres, and because *da-*22-to* is also in the inscription, Godart proposes that the royal personage referred to in the inscription was the ruler of the 'city' of Armenoi. He explains that production and export took place under orders of the wanax (*wa-na-ka-te-ro* = belonging to the king). This provides further evidence that the 'king' in the inscription cannot be the king of Knossos (as does the date of the vase).

Knossos tablets **KN C 979**, **KH Ar 2**, **KN C** point to the existence of 13 cities in West Crete. Tablet **C 979** refers to four related localities under the control of a provincial leader who was given the title of *a-to-mo*. Godart notes that, as a price for his administering the area, he received a pig. Godart points out that the words (singular or plural) *Wa-n-ka* (Head of State)/*Basileis, qa-si-re-we* (in the plural), in the Linear B tablets simply refers to the officials responsible for production of various commodities, not heads of state as in Homer's *Iliad* and *Odyssey*. The Knossos tablets record craftsmen, workshops for fabrics, furniture and precious objects directed by an official, a *Basileis* (scribe 102), at the 'city' of *da-*22-to*. Tzedakis has identified two pottery workshops at Armenoi (Appendix to Chapter 6). A pigment workshop was identified in the Bee Garden excavation (Fig 9.16). The tablets also make reference to agricultural products, spices and land at *da-*22-to*.

Godart identified in the **Co** series (scribe 107) a group of six known centres for cattle breeding as 'the block of the Cretan Far West'. He writes that it required fertile, well-irrigated land, which did exist in the larger area of the Necropolis (between the Necropolis and the sea) and names towns located between Chania and Souda Bay (Chapter 7). Two names in the text correspond to Chania and Aptera: *ku-do-ni-ja* and *a-pa-ta-wa*. Tablet **Ce 59**, from the Room of the Chariot Tablets at the palace of Knossos record oxen being reared in the region extending from the Messara plain (Amari Valley) to the outskirts of Rethymnon. In tablet **Ce 59.2b**, *da-*22-to* is listed as one of the areas known for cattle breeding. It is of particular interest that as early as Late Minoan II (in the **Ce 59** tablets, said to be Late Minoan II) at the time of the Mycenaean conquest of the island, West Crete was designated as an area particularly suited for raising cattle. Cattle bones were found at the Bee Garden during the 'city' excavation (see above and Chapter 8). Additional Linear B tablets from Knossos that include the place name *da-*22-to* as that of an important city, one with a king/leader, a centre of cattle breeding, agricultural, craft production, export/import, and support Godart's belief that *da-*22-to* is the 'city' of Armenoi that built the Necropolis of Armenoi are:

a) **AS 40**, Scribe 102, records an official(s) in charge of personnel and workshops (etc.) at *da-*22-to*
b) Series V, Scribe 125.
c) **V 756** and **1002**: *da-*22-ti-ja*, with feminine ending.
d) **X 7974**.

Tablets that hold references to shipping are:

a) **EL Z 1**.1: The export of a product contained in the stirrup jar discussed above, that dates to Late Minoan IIIB:1, was connected to *da-*22-to* (to 'Eleusis in particular') had been done under a palatial authority (see above).
b) **AS 40**, Scribe 125: records that *da-*22-to* controlled ships.
c) **Series V** and **X** name *da-*22-to*. The towns named in the series are in coastal areas and said to control sailors and ships.
d) **Series V 756** and **1002:** the tablets relate to 'transactions that concern the sea'. *da-*22-to* in this context has a feminine ethnic ending, *da-*22-ti-ja* which Godart accepts as referring either to the home port from which the sailors departed along the northern coast (an explanation he thought more credible) or to the name of a ship on which they served. That the 'city' which built the Necropolis produced many exportable products, e.g. ceramics, bronzes, soft stones, semi-precious stones, is documented. It was not far from the sea, and evidence has been found for a port at the end of a gorge from Somatas, a village just above the Necropolis, which extends from Gallos to the sea.

Conclusions

The village of Kastellos is situated not far from the north coast with a view of the sea. It is on the primary north–south route in the west of the island and enjoys a clear view of the passage After the Mycenaean invasion at the end of Late Minoan II, when the Minoan thalassocracy became the Mycenaean thalassocracy, the flow of trade reversed (became south to north) which meant a trading post at the northern end assumed an importance it had not previously enjoyed. In addition the area boasted mineral wealth. The village of Kastellos is known for its good climate. These are the reasons why a 'city' would have been founded where it was and why it would have succeeded and become a wealthy one. It would have been viewed in the era in which the 'city' was built as a prime location.

Discovery of a copper mine

In 2023, however, a new discovery was made which may indicate a further, possibly conclusive, reason why this location was chosen. This was the discovery of a copper mine

at Ano Valsamonero, 4 km from the Necropolis, an area also important for its iron deposits (Chapter 8; Sherwood-Dickinson *et al.* 2018). The mine is small but large enough to have been able to supply the needs of the bronze workshop at Armenoi (Tzedakis, pers. comm.). This on its own would have made the 'city' wealthy. Sherwood-Dickinson *et al.* (2018, 247) noted the presence of some poor quality and shapeless iron objects and one lump of lead and suggested that these represented early experiments in smelting iron and lead. Access to a copper source would have provided a vital ingredient for the manufacture of bronze and supports Giże's conclusion that the inhabitants obtained some of their wealth from metallurgy.

The significant aspects of the Linear B tablets that support the identification of Armenoi as *da-*22-to* can be summarised:

1) The geographical location (general) was between the Messara and the north coast of Crete;
2) the geographical location (specific) was in the vicinity of Rethymnon and not far from the sea;
3) the city controlled ships and shipping (general);
4) toponyms and toponymic derivatives painted on stirrup jars were found on the Mainland at Thebes;
5) Stirrup jars were exported from the 'city' both locally (Knossos) and to the Mainland (Eleusis; Midea). Some exports were made under the aegis of a 'king'. Some (not all) tin-plated goblets found in tombs were imported from the Mainland;
6) the presence of a royal tomb at the Necropolis, Tomb 159, indicates a wealthy town with a ruler;
7) goods were produced under the direction of a *basileis,* an official responsible for a group of craftsmen.

That there were various *basileis* in *da-*22-to* attests to craft production of the highest quality. According to Godart, this alone provides confirmation *da-*22-to* was the wealthy town of Armenoi, that built an outstanding necropolis.

Stable Isotope Analysis and DNA results indicate the presence of individuals who were born outside Crete and others who had spent parts of their lives outside Crete buried in the Necropolis, providing evidence for the cosmopolitan population of a wealthy 'city'.

5. Unlocked secrets

Chapters 2, 3, 4, 5, and 8 have unlocked secrets of those who were buried in the Necropolis. Chapters 6 and 9 combine to unlock further secrets, what the scientific results mean in terms of the society that built the wealthy Late Minoan III Necropolis of Armenoi, and the reasons why the wealthy polity that built it, must be the prominent city called *da-*22-to* in Linear B. An overview of the Necropolis as discussed in this chapter underpins Godart's conclusion that *da-*22-to* was the 'city' of Armenoi. In fact this discussion has turned up even stronger evidence than Linear B:

1. The scientific results reveal the tombs were multi-generational. They reveal a complex cosmopolitan society;
2. Ten per cent of the tombs are rich and 3% very rich. This is a very high percentage in general and certainly in any assemblage of Minoan chamber tombs. There were a surprising number of wealthy families living in Armenoi. Although there are tombs with small chambers and few grave goods, by far the majority, as evidenced by the finds, represent families who were fairly well off.
3. The existence of the Necropolis gives Godart's interpretation of the Linear B tablets all the support it needs even without the above summaries and the hugely important discovery of a copper source. The existence of a magnificent Necropolis is, in itself, a chief identifying factor that the 'city' of Armenoi was *da-*22-to.*

Questions

On consideration, could building the chamber tombs and producing what was found inside them, the goods and the trade that kept the 'city' prosperous, suggest a far larger population in and around the 'city' than those who were buried in the 182 confirmed chamber tombs (not counting the 50 enigmatic structures without chambers which are still under study). Could seven generations, and approximately 1000 people over a period of 200 years, have produced the grave goods and maintained the overall production of the 'city' – met its agricultural needs (sheep, goats, cattle and crops), and operated workshops for pottery, larnakes, bronzes, seals, etc, which kept the city alive, rich and thriving? The Knossos tablets record craftsmen, workshops for fabrics, furniture and precious objects directed by an official, a Basileis (scribe 102), at *da-*22-to.* Could only those who were buried in the Necropolis have produced all this? Was the population dependent on labour inside the 'city' and out, itinerant potters, perhaps, and others, to produce so much, as well as the building of farmhouses, townhouses and tombs? If so, who were they, how many were there, and how and where did they live?

Even the inhabitants of the poorer tombs do not represent the non-elite in society. There is a type of equality represented between the rich and the well-off and the not so well-off. It is clear that the less well-off still possessed some status in society. They had privilege: they had been recognised as making a valued contribution to society and because of that, had been granted the right to be buried in the Necropolis. Yes, there was a ruler in a royal tomb, and others around him who could display their wealth but what is indicated, in a highly unusual way, is that he/they were the 'first among equals'. The Late Minoan III Necropolis of Armenoi is, in its own distinctive way, a remarkable burial ground but begs a question still to be investigated. Where were all the others buried?

Appendix: catalogue

(original compilation by Vicky Kolivaki)

TOMB 159

Chronology: LM IIIA:1-IIIB:1

Contents of chamber

POTTERY

Cup

> R.M. 3457. Conical, tin-plated. H.: 3.5 cm. Workshop: Unknown. LM IIIA:2–LM IIIB:1.
> Incense burner. R.M. 3463. Base. H.: 7.5 cm. Workshop: Unknown. LM IIIB.

Kalathos

> R.M. 3456 H.: 11.2 cm. Workshop: Unknown. LM III A:2.

Kylikes

> R.M. 3455. Kantharoid, tin-plated. H.: 18 cm. Workshop: Knossos. LM III A:1.
> R.M. 3459. Loop-handled, tin-plated. H.: 11.8 cm. Workshop: Armenoi 1. LM III A:2.
> R.M. 3460. Kantharoid, tin-plated. H.: 20.1 cm. Workshop: Knossos. LM III A:1.

Stirrup jars

> R.M. 3447. Oval, decorated (Octopus FM. 21). H.: 32 cm. Workshop: Armenoi 1. LM IIIB:1.
> R.M. 3448. Piriform, decorated (Octopus FM. 21; chevrons FM. 58). H.: 41 cm. Workshop: Armenoi 1. LM IIIB:1.
> R.M. 3452. Squat, decorated (concentric semicircles FM. 43). H.: 14 cm. Workshop: Kydonia. LM IIIA:2–LM IIIB.

BRONZES

> R.M.M. 600. Knife. L: 23.6 cm. LM III:A2–LM IIIB.

LARNAKES

> R.M. 22506 body; R.M. 22507 lid (Larnax A). L.: 1.13–1.14 m; W.: 0.41–0.42 m; H.: 1.03 m with gable lid. Body: first long side: 2 zones separated by thin bands with lozenge chain FM.73; alternating arcs FM.69; upper zone hanging spirals FM.51:23; papyrus derivatives FM.11. Below the zones are horns of consecration FM.36 with double axe RM.35. Second long side: 2 uneven zones separated by thin bands: upper zone alternating arcs FM.69; below zone pendent rock pattern FM.32 with thin wavy band FM.53; horns of consecration FM.36 with double axe FM.35. First narrow side: 2 uneven zones separated by thin bands; upper zone lozenge chain FM.73; arcs; below zone pendent rock pattern FM.32 with thin wavy band FM53;

horns of consecration FM.36 with double axe FM.35; second narrow band: three uneven zones separated by think bands; upper zone lozenge FM.73; arcs; papyrus derivatives RM.11; middle zone concentric arcs FM.44:10; below papyrus derivatives FM.11. Lid: long and short sides: alternating arcs FM.69. LMIIIA:2/B:1.

R.M. 2690 (Larnax B). In fragments; at least 31 body fragments; 21 fragments of the lid. Body: L.: ca. 1.22–1.23 m; W.: ca. 0.51 m; H.: ca. 0.83–0.90 m. Lid: L.: ca. 1.28 m.; W.: ca. 0.56 m; H.: ca. 0.3 m. Body: long and short sides: wavy line FM.53:14. Lid: long sides and one short side alternating arcs RM. 69; second short side triangle FM.61A; bands. LMIIIA:2/B:1

R.M. Larnax 22505 The broken pieces were identified by Tzedakis and reported to the editors 11 July 2023. LMIIIA:2/B:1. A full description will be published in Volume III.

SMALL FINDS
Figurine (bust): ceramic protome
R.M. 9422 Head and part of neck. Broken nose and relief eyes added separately. Zig-zag (choker) painted on the neck and fully painted 'hat' on the head, which rises to a point in the centre and at the front. It features a large circle on the top with lines radiating out of it. H.: 7.0 cm; D.: 3 cm. LM IIIA:2.

Jewellery
R.M.M. 3241. Gold Pendant, leaf-shaped. L.: 3.0 cm; W.: 2.3 cm. LM IIIA

OTHER FINDS
R.M. ORG 01. Carbon, irregular shape. L.: 3.0 cm; W.: 2.3 cm.

R.M. ORG 02. Niche. Carbon, square. L.: 1.8 cm; W.: 1.5 cm.

R.M. ORG 03. Niche. Carbon, irregular shape. L.: 1.5 cm; W.: 0.8 cm.

R.M. ORG 04. Niche. Carbon. ca. 45 pieces, irregular shape. L.: 0.6–1.5 cm; W.: 0.5–1.2 cm.

Contents of dromos

POTTERY
Bowls
R.M. 17267. 2 sherds: base and everted rim. Pres. max. dim.: base Th.: 0.1 cm. LM IIIA:2.

R.M. 17667. 2 sherds: base and everted rim. P Pres. max. dim.: ex. H.: 7.0 cm. LM IIIA:2–LM IIIB.

Incense burner
R.M. 17272. 4 sherds: base and three feet join, cylindrical lid. Decorated (bands). Pres. max. dim.: base D.: 6.0 cm. LM IIIA:2–LM IIIB.

Kraters
R.M. 17246. Amphoroid. 5 sherds: shoulder and neck join. Decorated (octopus FM.21). Pres. max. dim.: wall Th.: 0.6 cm. Workshop: Armenoi 2. LM IIIA:2.

R.M. 17266. Amphoroid. 1 sherd: rim and strap handle join. Decorated (bands; triangles FM.61A). Pres. max. dim.: handle Th.: 4.3 cm; rim Th.: 1.5 cm. LM IIIA:2.

R.M. 17269. Amphoroid. 2 body sherds. Decorated (octopus FM.21). Pres. max. dim.: 6.5–8 cm. Workshop: Armenoi 1. LM IIIA–LM IIIA:2.

R.M. 17275. Amphoroid. 1 sherd: neck and everted rim join. Decorated (bands). Pres. max. dim.: ex. H.: 11 cm. Workshop: Armenoi 1. LM IIIB1 end.

R.M. 17276. Amphoroid. 2 sherds: neck everted rim and strap handle. Decorated (semicircles FM.43; bands). Pres. max. dim.: ex. H.: 20 cm. Workshop: Armenoi 1. LM IIIA:2.

R.M. 17674. Amphoroid. 7 sherds: wall, neck and strap handle. Decorated (octopus FM.21; quirk FM.48). Pres. max. dim.: handle Th.: 4.0 cm. LM IIIA:2–LM IIIB.

Kylikes
R.M. 17249. 2 sherds: stem and disc base. Pres. max. dim.: stem H.: 6.5 cm; stem D.: 2.3 cm. LM IIIA:1

R.M. 17268. 3 sherds: disc base, strap handle and wall sherds. Decorated (bands) Pres. max. dim.: handle Th.: 1 cm. Workshop: Knossos. LM IIIA:1

Jugs
R.M. 17248. 1 sherd: wall and strap handle join. Decorated (bands). Present max. dim.: handle length: 6.5 cm. LM IIIA:2.

R.M. 17271. 1 sherd: part of neck and everted rim join. Decorated (remains of paint). Pres. max. dim.: ex. H.: 3.3 cm. LM IIIA:2.

R.M. 17273. Bridge-spouted. 3 sherds: spout and wall. Decorated (bands; octopus FM.21). Pres. max. dim.: wall Th.: 0.8 cm. Workshop: Armenoi 1. LM IIIB:1.

Pithos
R.M. 17270. 2 body sherds. Decorated (relief bands). Present max. dim.: wall Th.: 1.5 cm. LM IIIA:2 end.

Skyphos
R.M. 17257. 1 sherd: body, everted rim and strap handle join. Present max. dim.: ex. H.: 5.3 cm; handle L.: 5.5 cm. LM IIIA:2.

Stirrup jars

R.M. 17246. 1 sherd: ring base and wall join. Present max. dim.: base D.: 12 cm; ex. H.: 4.5 cm. Workshop: East Crete. LM IIIA:2–LM IIIB.

R.M. 17265. A: Spout and shoulder join. Decorated (bands). Pres. max. dim.: ex. H.: 3.5 cm. B: Spout. Decorated (bands). Present max. dim.: ex. H.: 2.7 cm. Workshop: Armenoi 1. LM IIIA:2.

TOMB 146

Chronology: LM IIIA–LM IIIB

Contents of chamber

POTTERY

Incense burner

R.M. 3405. Only the lid, decorated (lozenge FM.73). H.: 11.5 cm. Workshop: Knossos. LM IIIB.

Stirrup jars

R.M. 3363. Unique vase. Linear B inscription: *wi-ni-jo*. Piriform, decorated (bands). H.: 31.3 cm. Workshop: Armenoi. LM IIIA:2–IIIB:1.

R.M. 3410. Squat, new-made, decorated (bands/traces). H.: 12.0 cm. Workshop: Armenoi. LM IIIB.

SHERDS

Stirrup jars

R.M. 17233. ca 49 sherds from piriform jar: ring base, wall, false spout and vertical handles. Decorated (bands). Pres. max. dim.: base D.: 5.5 cm; spout D.: 3.1 cm. Workshop: Kydonia. LM IIIA:2.

R.M. 17236. 13 sherds from miniature vase: ring base wall, vertical strap handles and false spout. Decorated (bands, curved lines). Pres. max. dim.: wall Th.: 0.3 cm; spout D.: 2.0 cm. Workshop: Kydonia. LM IIIB.

SEAL

R.M. Σ126 (*CMS* V Suppl. 1B no. 271). Lentoid. Serpentine (?). Heavily abraded; edges battered; String holes worn. D.: 2.08–2.11 cm. Motif virtually illegible: 2 recumbent quadrupeds (?) Late Minoan.

LARNAX

R.M. 17209. 2 sherds which join. Fully painted rim. Stylised octopus tentacles on body. (Octopus, FM.21). LM III.

Contents of dromos

POTTERY

Alabastra

R.M. 17240. 2 sherds: neck and everted rim, decorated (bands, concentric arcs). Pres. max. dim.: ex. H.: 10.0 cm, wall Th.: 0.6 cm. LM IIIA:2/B1.

Bowls

R.M. 17154. 1 body sherd, decorated (bands). Pres. max. dim.: wall Th.: 0.4 cm. LM IIIA.

Cups

R.M. 17163. 1 sherd: flat base with hole and wall, decorated (paint remains). Pres. max. dim.: ex. H.: 6.0 cm. LM IIIA.

R.M. 17235. 2 sherds: wall and rim, decorated (flowers). Pres. max. dim.: wall Th.: 0.4 cm. LM IIIA.

Cooking pots

R.M. 17149. 6 sherds: feet and horizontal handle. Pres. max. dim.: H.: 9.0 cm. LM IIIA.

R.M. 17157. 1 sherd: wall and horizontal handle join. Pres. max. dim.: wall Th.: 0.8 cm. LM IIIA.

R.M. 17158. 17 sherds: flat base, feet, wall, tree handles and everted rim. Pres. max. dim.: base Th.: 0.1 cm, foot height: 13.0 cm. LM IIIB.

Kraters

R.M. 17159. Amphoroid. 1 sherd: neck and everted rim join. Decorated (paint remains inside/outside). Pres. max. dim.: wall Th.: 0.5 cm. LM IIIA2/B1.

Kylikes

R.M. 17151. Kantharoid. 3 sherds: body and vertical strap handle, decorated (fully painted inside/outside). Pres. max. dim.: wall Th.: 0.3 cm. LM IIIA.

R.M. 17153. 1 sherd: stem and wall join. Pres. max. dim.: ex. H.: 10.0 cm.

R.M. 17162. 1 sherd: stem and wall join. Decorated (whitish slip) Pres. max. dim.: ex. H.: 14.0 cm. LM IIIA2.

R.M. 17164. 2 sherds: wall, everted rim and vertical strap handle. Decorated (paint remains). Pres. max. dim.: wall Th.: 0.3 cm. LM IIIA.

R.M. 17168. 1 sherd: disc base and stem. Decorated (paint remains) Pres. max. dim.: ex. H.: 7.5 cm. LM IIIA.

R.M. 17207. 1 sherd: disc base, stem and wall sherd. Decorated (paint remains) Pres. max. dim.: ex. H.: 12.0 cm.

R.M. 17208. 1 sherd: disc base, part of stem and wall sherd. Decorated (fully painted inside). Pres. max. dim.: ex. H.: 5.0 cm. LM IIIA2.

R.M. 17216. 1 sherd: wall, everted rim and strap handle. Decorated (paint remains). Pres. max. dim.: wall Th.: 0.4 cm. LM IIIA.

R.M. 17218. 5 sherds: disc base and wall sherds. Decorated (paint remains inside). Pres. max. dim.: wall Th.: 0.5 cm. LM IIIA.

R.M. 17237. Kantharoid, tin-plated. 2 sherds: strap handle, everted rim and wall sherds. Decorated ('fully painted'). Pres. max. dim.: wall Th.: 0.3 cm. LM IIIA:2.

R.M. 17238. 10 sherds: disc base, strap handle, everted rim and wall sherds. Decorated (paint remains) Pres. max. dim.: wall Th.: 0.4 cm. LM IIIA:2.

R.M. 17243. 9 sherds: disc base, stem, strap handle, and wall sherds. Decorated (fully painted inside, bands outside) Pres. max. dim.: wall Th.: 0.4 cm. LM IIIA:2.

R.M. 17244. 2 sherds: stem and strap handle. Decorated (bands) Pres. max. dim.: wall Th.: 0.4 cm. LM IIIA:2.

Pithos

R.M. 17215. 7 body sherds. Decorated (rope-shaped bands and small relief bands). Pres. max. dim.: wall Th.: 1.0 cm. LM IIIA:2 end.

Stirrup jars

R.M. 17150. 1 sherd: ring base and wall join. Decorated (band). Pres. max. dim.: wall Th: 1.0 cm. LM IIIA:2/IIIB.

R.M. 17210. 10 sherds. Flat base and wall sherds. Decorated (bands). Pres. max. dim.: base D.: 12.0 cm. Workshop: East Crete. LM IIIB.

R.M. 17211. 2 wall sherds. Decorated (Octopus, FM.21). Pres. max. dim.: wall Th.: 0.6 cm. LM IIIA:2.

R.M. 17234. 4 sherds. Flat base. Decorated (paint remains). Pres. max. dim.: base D.: 12.0 cm.. LM IIIA.

R.M. 17239. 1 sherd. False spout, shoulder and vertical strap handles join. Pres. max. dim.: wall Th.: 0.5 cm. LM IIIA:2/B1.

Note

1   The Surveys were funded by The Headley Trust, INSTAP, and The Holley Martlew Archaeological Foundation. The Holley Martlew Foundation funded the excavations at Armia and the Bee Garden. For acknowledgement see Chappell and Allender (2018); Tzedakis *et al.* (2018, xix–xxi).

Bibliography

Ariotti, A. (2018) Roman and Early Byzantine diagnostic pottery from the field surveys: 2001, 2002 and 2007. In Tzedakis *et al.* (eds), 111–212.

Barrington, G. and Chapman, C.E. (2004) A high-stability fluxgate magnetic gradiometer for shallow geophysical survey applications. *Archaeological Prospection,* 11, 19–34.

Catling, H.W., Cherry, J.F., Jones, R.E. and Killen, J.T. (1975) The Linear B inscribed stirrup-jars and western Crete. *Annual of the British School at Athens* 75, 49–113.

Chappell, E. and Allender, S. (2018) Site investigations of the Necropolis and its environs: the search for the town. In Tzedakis *et al.* (eds), 23–52.

Giże, A.P. (2018a) Topographical setting. In Tzedakis *et al.* (eds), 19–22.

Giże, A.P. (2018b) Geological setting. In Tzedakis *et al.* (eds), 213–30.

Giże, A.P. (2018c) Proposed method of tomb construction. In Tzedakis *et al.* (eds), 231–40.

Kanta, A. (1980) *The Late Minoan III Period in Crete: a survey of sites, pottery and their distribution.* Gothenberg, Studies in Mediterranean Archaeology 58.

Krzyszkowska, O. (2005) *Aegean Seals: An Introduction (BICS* Suppl. 85). London.

Krzyszkowska, O. (2019) Changing perceptions of the past: the role of antique seals in Minoan Crete. In E. Borgna, I. Caloi, F.M. Carinci and R. Laffineur (eds), *MNHMH/ MNEME. Past and Memory in the Aegean Bronze Age.* Proceedings of the 17th International Aegean Conference. University of Udine, Department of Humanities and Cultural Heritage, Ca'Foscari University of Venice, Department of Humanities, 17–21 April 2018. Aegaeum 43 (Peeters: Leuven – Liège) 487–496.

Martlew, H. forthcoming. The Unfinished Tombs in the Late Minoan III Necropolis of Armenoi. In M. Andreadaki-Vlasaki and I. Gavrolaki (eds), *Festschrift in Honour of Yannis Tzedakis.* Rethymnon, Ephorate of Antiquities.

Martlew, H., Giże, A.P. and Kolivaki, V. (2018) Minoan diagnostic pottery from the field and geophysical surveys: 1992, 1997, 2001, 2002 and 2007. In Tzedakis *et al.* (eds), 67–110.

Masters, P. (2018) Geophysical survey: Necropolis and town. In Tzedakis *et al.* (eds), 53–65.

Peatfield, A. (1994) The Atsipadhes Korakias Peak Sanctuary Project. *Classics Ireland* 1, 90–5.

Sherwood-Dickinson, C., Droop, G. and Giże, A.P. (2018) Ano Valsamonero iron deposit: a potential metal source for the Late Minoan III community. In Tzedakis *et al.* (eds), 241–8.

Tzedakis, Y. and Kolivaki, V. (2016) The Minoan Polis and the Necropolis of Armenoi. In P. Karanastasi (ed.), *Archaeological Work in Crete. Proceedings of the 4th meeting Rethymnon, 24–27 November 2016* II, 595–601. Rethymnon, Ephorate of Antiquities.

Tzedakis, Y. and Kolivaki, V. (2018) Background and history of the excavation. In Tzedakis *et al.* (eds), 1–18.

Tzedakis, Y. and Martlew, H. (eds) (2003) *Minoans and Mycenaeans: flavours of their time* (revised edition). Athens, Ministry of Culture.

Tzedakis, Y., Martlew, H. and Jones, M.K. (eds) (2008) *Archaeology Meets Science: biomolecular investigations in Bronze Age Greece.* Oxford, Oxbow Books.

Tzedakis, Y., Martlew, H. and Arnott, R. (eds) (2018) *The Late Minoan III Necropolis of Armenoi,* Volume I. Philadelphia PA, INSTAP Academic Press.

10

Western Crete, the 'city' of Armenoi and the fall of Pylos

Louis Godart

Introduction

On the dating of the fall of Pylos, Carl W. Blegen had no doubt:

> The exact date of the destruction can hardly be fixed to within a year, but it came when the Mycenaean ceramic of the style named IIIB was coming to an end and a number of vases of the following style, the Mycenaean IIIC, was beginning to appear. It was a time of turmoil and destruction. Mycenae and Tiryns in turn were destroyed by fire when ceramics of style IIIB were close to disappearing. Many other Mycenaean sites such as Berbati, the Heraion of Argos, Zygouries, Thebes and Gla to mention a few, disappeared at the same time. (Blegen and Rawson 1967, 32)

Today most of those who excavate in Pylos follow the Master and date the end of the palace of Nestor to the Late Helladic IIIB2 or the very beginning of the Late Helladic IIIC, that is to say, at the extreme end of the 13th or at the very beginning of the 12th century, between 1200 and 1180 BC. Things are, however, nowhere near that simple.

The images that lie dormant in our minds when we evoke the Mycenaean palaces of mainland Greece are those of Mycenae, the proud citadel that dominates the lands of Argos, the fortress of Tiryns close to the sea, Midea perched on its mountain, Thebes with the seven gates over which Cadmos reigned. All these princely residences have a common characteristic: they are defended by imposing walls.

The history of the construction of these fortifications is spread over three phases throughout the 14th and 13th century. In Tiryns a first fortification was built at the beginning of the Late Helladic IIIA (1370 BC). Then, around 1300, there is a first phase of extension that strengthens the defence of the citadel by modifying the entrance system to the east and south and by enlarging it by the development to the north of an additional space, the middle citadel. It was then that the palace was built, which was destroyed around 1250.

Finally, a third phase of extension led to the construction of the casemates to the south, the enormous western bastion and the establishment of the lower citadel.

At Mycenae it is also observed that the walls were built in three phases. A first enclosure was built at LH IIIA2 (around 1340 BC) at the top of the acropolis, where the palace was built. After the destructions of 1250, an extension to the south was made to include Circle A, excavated by Heinrich Schliemann in 1876, and the famous 'Lion Gate' was built. Finally, shortly before the final destruction around 1200 BC, there was another extension to the north-east. Two galleries were pierced in the walls; one to the south with a corbelled vault led to a small lower terrace, the other to the north was intended to protect access to the underground cistern of the acropolis built around 1300 BC and destroyed in the second half of the 13th century.

Two other Mycenaean palaces have been unearthed on the Peloponnese mainland: Pylos in Messinia, excavated by Carl W. Blegen in 1939 and again from 1952 to 1966, and the palace of Aghios Vasileios in Laconia that has been excavated since 2009 by a Greek team led by Adamantia Vassilogamvrou. The physiognomy of these last two palaces contrasts sharply with that of other continental Mycenaean palaces: no cyclopean apparatus around what was the palace of old Nestor or the residence of the blond Menelaus. What for? Is it reasonable to suppose that, as the vulgate teaches, all these constructions are more or less contemporaneous and that the destruction of these sites occurred at the same time?[1]

Mervyn Popham (1991) asked himself the question about Pylos, although he did not know the excavation of Aghios Vassileios: 'Why the palace at Pylos was unfortified at the time of its destruction, which was usually placed at the very end of the century [the 13th century BC]' (Popham 1991, 315). In his article Popham demonstrates that there was a re-occupation of the site of Pylos after the fall of the Mycenaean palace because vases from the Late Helladic IIIC, or the very beginning of the Iron Age, were erroneously attributed to the final catastrophe which, in reality as the re-examination of ceramics suggests, dates back to the end of Late Helladic IIIA2 or the very beginning of Late Helladic IIIB1 (1300 BC). This same problem of the absence of fortifications at Pylos intrigued Jean-Claude Poursat who writes (2014, 147):

> We wondered what needs corresponded to the construction of these fortifications, at a time when nothing seems to threaten the prosperity of the sites concerned. However, if we adopt the high date of a destruction of the palace of Pylos around 1300, it is from this period that turbulence emerges in Greece.

Moreover, it was the time when the city of Bogazköy in Anatolia was surrounded by an enlarged enclosure: its fortifications included glacis and walls, towers, posterns similar to those of Tiryns, impressive gates (Bittel 1976, fig. 101).

P.M. Thomas, in turn, faced the question of the dating of Pylos (2004, 207–24). If, Thomas writes, the palace was in use throughout the period of Late Helladic IIIB, where are the Group B of the deep bowls, the deep rosette bowls and the undecorated conical kylix so characteristic of the Late Helladic IIIB2 period (1250–1200 BC) in Argolis, Corinth and Attica? Popham's suggestion that the palace of Pylos was destroyed at the extreme beginning of Late Helladic IIIB, perhaps shortly after the beginning of Late Helladic IIIB1 (1300 BC), deserves further consideration. If we place the destruction of Pylos at the beginning of Late Helladic IIIB, we solve the question of the dating of most of the ceramics discovered in the palace as well as the absence of typical forms of Late Helladic IIIB2 and we manage to explain the significant difference between the ceramics that appeared at sites such as Nichoria and the ceramics of Argolis. If the palace had been in operation throughout the period of Late Helladic IIIB, why would the close historical relations with Argolis in the field of ceramics and many others have been interrupted?

Bartłomiej Lis published an article in 2016 questioning the dating of the fall of Pylos (2016, 491–536). At the end of his examination, Lis points out that the group of vases, partly modelled by hand, from Room 60 contrasts with the whole of Pylian ceramics. At this point, he adds, one must recall the minute amount of decorated ceramics, the small number of vases found in rooms other than kitchens, and the mass of stylistically older vases belonging to a period prior to the date generally proposed for the fall of Pylos, that is, around 1200 (Lis 2016, 532–3).

In a letter, Poursat wrote to the present author:

> There was very recently, on 8 and 9 November in Vienna, a 'workshop' entitled 'Synchronising the Destructions of the Mycenaean Palaces'. In the summary of their paper, J. Davis and his colleagues maintain for the destruction of the palace of Pylos a date at the very beginning of the IIIC

(with a first destruction of the palace at the beginning of the IIIA2 [between 1350 and 1320]), contesting and minimising the importance of ceramic analyses such as that of P. Thomas. In the conversations I had, however, I noted that, for the nearby site of Aghios Vasileios, the violent destruction of the site (with the tablets) would take place at the end of HR IIIB1, around 1250, with final abandonment of the site around 1200. Would there have been destruction at Pylos around the same date as well? (Poursat, pers. comm. 8 December 2016)

Seals and seal impressions

The analysis of seals and seal impressions discovered both in the tombs of Messinia near the palace of Pylos and in the layers of destruction of the building itself provides elements worthy of inclusion in the debate on the dating of the fall of Pylos. I will quote Poursat (2014, 112; translated):

> Many sealings found in the destruction layer of the Palace of Pylos (Late Helladic IIIB) were impressed by metal signet rings, identifiable by the convex oval shape of the imprints they made. Stylistically, these rings have been dated to Late Helladic IIB–IIIA1 (1450–1370 B.C.).

Their imprints enrich our knowledge of the iconographic repertoire of this period: scenes of struggle between men and lions, men and griffins, are added to the animal fights mentioned above (Poursat 2014, figs 120–1). Other representations show griffins in heraldic position, with a wide spiral on the chest (fig. 120); their attitude evokes the griffins of the throne rooms of Knossos and Pylos, and their style of the ivory plates of the following period (fig. 231). A frieze of dogs and griffins, surmounting a frieze of double argonauts from which it is separated by a band of chained spirals (Fig. 124), is directly comparable to the murals of the palace of Pylos where we will find these elements.

It is ironic that most commentators, with the exception of Poursat, consider that all users of Pylos seals would have inherited stamps dating back more than 150 or even 200 years. It is certainly possible that in some cases at least, these seals come from inheritances that would have crossed the centuries (from 1450/1370 to 1200/1180 BC) if we place the end of Pylos in Late Helladic IIIB2/LH IIIC) but it is more difficult to admit that *all* the seals used in Pylos belonged to the distant ancestors of the administrators of the palace if we place the end of the latter between 1200 and 1180 BC. Poursat (2014, 178; translated) states that:

> In mainland Greece sealings are numerous in LH IIIB contexts at Mycenae, Pylos, Thebes, Tiryns, but everywhere the imprints seem to be from antique seals kept by officials of the administration. At the palace of Pylos most of the imprints match seals of Late Helladic IIB–IIIA1.

In addition, we note with Poursat, the narrow parallelisms between the representations incised in the seals in question and the frescoes (griffins in heraldic position for Knossos and Pylos; frieze of dogs and griffins surmounting a frieze of double argonauts from which it is separated by a band of chained spirals for Pylos). These elements strongly argue in favour of the hypothesis associating the end of Pylos with a time well before the Late Helladic IIIB2/IIIC.

The frescoes of Pylos

The decoration of the throne room of the palace of Nestor has obvious similarities with the decoration of the throne room of the Mycenaean palace of Knossos. It is enough to quote C.W. Blegen aptly insisting on the 'close ideological relationship associating these two Mycenaean compositions' with regard to the frescoes discovered in the throne room of Pylos (Blegen and Rawson 1966, 79 and fig. 74). Poursat, in turn, is struck by the similarity between the throne rooms of Knossos and Pylos and the wingless griffins in heraldic position around the throne (Poursat 2014, 173). Many studies have been devoted to this aspect of the relationship between Knossos and Pylos (Shank 2007). Stephan Hiller in particular raises the question of the relationship between the two palaces (Hiller 1966).

Ivories

Poursat, in his analysis of Mycenaean ivories, shows that it was at the time between Late Helladic IIB and Late Helladic IIIA1 (1450–1400 BC) that the repertoire of Mycenaean ivories was established (Poursat 2014, 117). Among the secondary elements of ivory decoration, Poursat cites the 'beading and grooving' found in Pylos, Argolis and Crete (Poursat 1977, 184–5). However, this type of decoration could be limited in time because the Katsambas comb dates from Late Helladic II or the beginning of Late Helladic III. Granulation in goldsmithing, adds Poursat, from which this motif is directly inspired, is a characteristic feature of Late Helladic II (Poursat 1977, 185). The ivories of Pylos are therefore related to a period very much earlier than Late Helladic IIIB2/IIIC (1200/1180 BC). However, as Poursat always notes, in the ivories of Pylos we find elements such as meanders, unknown on other sites, as well as a very particular treatment of *foliate bands* that recall the peculiarities that Mabel Lang was able to detect in the secondary decoration of the frescoes (Poursat 1977, 175–6). It is therefore more than tempting to associate the date of the frescoes of Pylos with that of the ivories presenting this type of decoration. In addition, in a letter, J.-C. Poursat points out to me that:

> the material decorated with ivory described in the tablets of the **Ta** series of Pylos would correspond better to a destruction at the end of LH IIIB1: it recalls the ivories found in Mycenae in the House of the Sphinxes or that of the Shields, or in the tombs of Spata, Phylaki or Chania, i.e. in LH IIIA2/IIIB1 contexts. At LH IIIB2, ivory seems much rarer in Crete as in Argolis. (Poursat, pers. comm. 8 December 2018)

Linear B documents and their chronology

Do the inscriptions in Linear B on vases on the one hand, and the tablets of Pylos on the other, provide us with data likely to bring new elements to the debate on the chronology of the palace of Nestor? Pylos had relations with Crete and Cretan craftsmen were present in Messinia. The tablets of the **Ta** series of Pylos stand an inventory of royal furniture. In **Ta 641.1** are mentioned two tripods (*ti-ri-po-de* *201^{vas}) a_3-*ke-u* (decorated with goats or provided with handles in goats' heads (Chantraine 1968, *S.V.* αἴξ) called *ke-re-si-jo*, *we-ke* κρησιο-ϝεργής 'of Cretan manufacture' and another tripod (*ti-ri-po*), in turn *ke-re-si-jo, we-ke*, whose feet are burned: *a-pu, ke-ka-u-me-no ke-re-a₂* *ἀπυκεκαυμένος *σκέλεhα (plural of σκέλος 'the foot') (Ventris and Chadwick 1973, 533–53. In **Ta 709.3** appear a tripod decorated with goats (or provided with handles in goats' heads) and a tripod whose decoration is described by the adjective *o-pi-ke-wi-ri-je-u* of difficult interpretation (Aura Jorro 1993, *S.V.*), both again qualified as *ke-re-si-jo, we-ke*, that is to say of 'Cretan manufacture'.

A tripod vase from a tomb of Volimidia, decorated with three animal heads (two deer, one bull) provides a good illustration of what scribe 602 describes in **Ta 641.1** (Poursat 2014, 212, fig. 292). This vase is dated by Robert Koehl (2016, nos 44–61) to Late Helladic IIIB1. Vases with animal heads are rare but there is a series in the workshop of Kydônia dating from the Late Helladic IIIA2 (1370–1300 BC) which corresponds and corroborates the mention of the 'Cretan manufacture' indicated in the tablet (Godart and Tzedakis 1992, 35–6).[2] All these vases appear to pre-date 1250 or even 1300. If, as Poursat writes to me (pers. comm. 8 December 2018), 'the scribe 602, author of the **Ta** series, recorded material of his time, the latter could in no way be later than LH IIIB1'. I add that the comparisons between the descriptions of scribe 602 in **Ta**, the vases with animal's heads from the workshop of Kydonia and the rhyton of Volimidia, argue in favour of a dating located at the hinge between the end of the 14th and the beginning of the 13th centuries for the archives of Pylos (between the end of Late Helladic IIIA2 and the beginning of Late Helladic IIIB1). This conclusion is consistent with that of Popham who, based on the re-examination of Pylos ceramics, suggests that the final catastrophe that swept away the palace dates back to the period between the end of Late Helladic IIIA2 and the beginning of Late Helladic IIIB1 (Popham 1991, 315–24).

In addition at Pylos (tablet **PY An 128,3**) mentions five *ke-re-te*, that is to say five Cretans (*ke-re-te* = Κρῆτες, plural of Κρής) called *ka-si-ko-no*, a term that undoubtedly serves to designate craftsmen.[3] The existence of itinerant craftsmen is well documented in the Mycenaean world; it is therefore likely that five Cretans exercising the profession of *ka-si-ko-no* were inserted into the mechanisms of the Pylian administration (Godart 2020, 211, 253).[4] The economic and cultural relations between Crete and the palace of Pylos were therefore on the agenda at the time of the tablets in Linear B discovered in the palace of Nestor.

At the time of the Late Minoan IIIB1 (1250 BC) the political and administrative centre of Κυδωνία in western Crete (the *ku-do-ni-ja* of tablets in Linear B) imposed itself in the big island after the fall of Knossos. This centre exported stirrup jars to the Greek mainland, especially to Mycenae, Tiryns, Midea, Thebes and Eleusis. The clay of these vases is indeed a clay from western Crete and the scribes who painted the texts covering the belly of these amphorae exercised their activity in Κυδωνία. Inscriptions painted by these scribes have been unearthed both in Κυδωνία and in the localities and continental Mycenaean palaces mentioned above (Godart and Sacconi 2017).

Is it not strange that no inscribed stirrup jars from western Crete have been discovered at Pylos when we know that the great Mycenaean centres of the continent have restored in abundance this kind of material and that Pylos, as evidenced by the tablets from the palace, had close relations with Crete? Wouldn't it be simply because the palace had ceased to exist *before* 1250, when the exports to the continent of stirrup jars from western Crete are located? If this is indeed so, it would be Knossos who would have been at the centre of relations with Messinia, of which the tablets in Pylos in Linear B bear witness and this at a time before the end of LH IIIB1.[5]

Tablets Ae 995, La 994, Xa 1419, Xa 1420 and Xn 1449

About these documents Palaima writes (1988, 164–5). 'Two other tablets from the SW Area were found in a different context and are disassociated not only from the mainstream of scribal activity in the palace proper, but even from the 34 tablets that constitute the primary data from the record-keeping in the SW Building. During the 1960 season, George Papathanasopoloulos excavated along and above the outer south-western wall of the SW Building. Unfortunately, his trench notebook is missing from the University of Cincinnati Excavations Archives but, according to C.W. Blegen, Papathanasopoulos began at the extreme western corner of Room 81 (marked GP 1 in Palaima 1988, 160, fig. 21) and moved south-eastward in increments of roughly 2 m until he eventually reached the main hall, Room 65, in GP 10. In Trench GP 2 he discovered tablet **Xa 1420** (Blegen and Rawson 1966, 20–1); in Trench GP 5, 0.85 m below the surface in a disturbed stratum with mixed stratigraphy, tablet **Xa 1419** (Blegen and Rawson 1966, 28–5). As noted in the analysis of hands, these tablets together with **Ae 995**, **Xn 1449** and **Ua 994** (today **La 994**) form a unique group, not only for Pylos but for the entire Greek mainland, possessing a distinctively Knossian-Cretan graphic style.[6] Running north-eastward inside the outer wall of the SW

Building from Room 65 into Room 81 is a segment of impressive wall antedating the Late Helladic IIIB palace and destroyed by the bedding trench for the outer wall of the SW Building. It is possible that **Xa 1419** and **1420** may be chance remains from the earlier structure deposited in the bedding trench when the SW Building was constructed (Blegen and Rawson 1966, 282–3). The palaeographically related tablets **Ua (La) 994** and **Ae 995**, are also from a context in which rooms of the later palace were built over successive phases of wall from the period LH IIIA or earlier (Blegen *et al* 1973, 35–7).

Tablets **La 994** and **Ae 995** were discovered in the area below Rooms 55, 56, 57 and the Archives Room (Blegen *et al.* 1973, 35–7). Palaima writes (1988, 169):

> In 1953 Theokaris, whose notebook is missing from the University of Cincinnati Excavations Archives, excavated Rooms 55–57. He uncovered, despite the confusion created by earlier intrusions, at least three successive and stratigraphically complicated phases of occupation, together with wares of Mycenaean IIIA (Blegen *et al.* 1973: 35–37). One example of the earlier period is a large pithos containing conical cups beneath the floor of Room 55 and belonging to a stage before the palace was built (Blegen and Rawson 1966, 223).

From these Rooms comes **Ae 995** (Fig. 10.1) and perhaps **Ua (La) 994** (Fig. 10.2).

Thus the group of 4 tablets coming both from the South West complex (SW Building), in particular from the area west of Room 81 (**Xa 1419 and 1420**; Figs 10.3–10.5)[7], as well as Rooms 55–7 located in the opposite region, the South-East wing of the palace (**Ae 995 and La 994**), dates to Late Helladic IIIA (ca. 1370–1320). As these are documents fired by chance in the fire that devastated these rooms, it is obvious that the palace of Pylos suffered destruction during the 14th century (Del Freo 2016, 194 and n 26) and probably towards the end of it. As Poursat has noted, even today, the excavators of Pylos agree on the fact that the palace suffered destruction at Late Helladic IIIA2, between 1350 and 1320 BC (Poursat, pers. comm. 8 January 2018). The tablet **La 994** has the base of a mutilated sign (the upper part has disappeared) whose identification is however certain: it is the logogram of the wool. This logogram, which evokes from afar a cat's head and which goes back to a well-known archetype in Linear A, has here a very particular characteristic: two eyes are drawn on either side of the shaft representing the nose of the animal (Ventris and Chadwick 1973, 314).

However, two tablets of Pylos, **La 632** and **La 635** (Figs 10.6 and 10.7) present the logogram of wool having these characteristics, namely the two eyes drawn on either side of the shaft that represents the cat's nose. Such a characteristic element is found neither in Pylos, nor in all the many texts in Linear B presenting the logogram of wool.

*Figure 10.1. Linear B Tablet **Ae 995** (scribe 691).*

 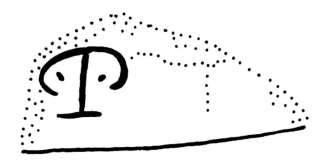

*Figure 10.2. Linear B Tablet **La 994** (Rooms 55 and 57?).*

*Figure 10.3. Linear B Tablet **Xa 1419** (south-west sector GP 5; scribe 691).*

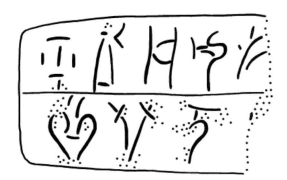 

*Figure 10.4. Linear B Tablet **Xa 1419**.*

 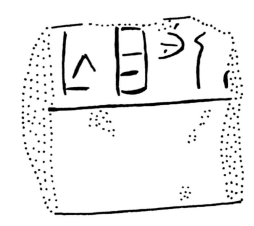

*Figure 10.5. Linear B Tablet **Xa 1420** (south-west sector GP 2: –).*

*Figure 10.6. Linear B Tablet **La 632** (Room 6, north-east sector).*

*Figure 10.7. Linear B Tablet **La 635** (Room 6, north-east sector).*

Here are some of the wool logograms attested in Pylos:

A 267.8 (scribe 601);   **A 443.2** (scribe 606)

The conclusion is obvious: the **tablets La 994, La 632** and **La 635** are from the hand of the same scribe. The tablets **La 632** and **La 635** are from Room 6 (Throne Room), North-east sector (Palaima 1988, 137). These documents were unearthed in 1952. They were attributed to the destruction layer of the palace, the one that Carl Blegen dates at Late Helladic IIIB2/IIIC (1200/1180 BC; Shelmerdine 1998, 294). But if, as we have seen above, **La 994** belongs to the period of Late Helladic IIIA2 (1320 BC), it is impossible that the documents **La 632** and **La 635**, which are by the hand of the same scribe, date from Late Helladic IIIB2/IIIC (1200/1180). More than a hundred years separate Late Helladic IIIA2 from Late Helladic IIIB2/IIIC and far exceeds the life expectancy of a Mycenaean scribe. It must therefore be concluded that the destruction of the palace of Pylos, as argued by Popham, Poursat, Thomas and Lis, dates back to a period well before

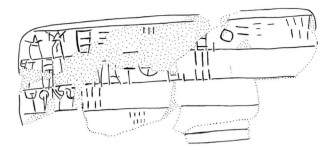

*Figure 10.8. Linear B Tablet **Vn 1339** to which fragment Xn 1449 was connected on the lower left.*

1200 BC. It is to this conclusion about **La 632** and **La 635** that Skelton arrives in two articles questioning the dating of the tablets discovered in the Megaron (or Throne Room) of the palace of Pylos (Skelton 2009; 2011). Doubts about the dating of the **La** tablets had already been put forward by Melena (2000–2001). These are weighty arguments to be added to the case relating to the destruction of Nestor's palace.

Xn 1449

This document **Xn 1449** (Fig. 10.8) that Palaima and the excavators of Pylos date from LH IIIA2 was connected to tablet **Vn 1339** which would go back to Late Helladic IIIB2/IIIC. The writing of this tablet, as Palaima argued, is to be compared to the writing of scribe 691 who wrote tablets **Ae 995** and **Xa 1419**, which in turn date from an earlier era (Late Helladic IIIA2). If this is indeed the case, it must be concluded that tablet **Vn 1339** discovered in Room 99 is also dated from Late Helladic IIIA2. Of course, the same is true for all documents associated with the area 99.

However, as it is said in Palaima's *The Scribes of Pylos* (Palaima 1988), in this Room 99 were discovered documents from the hand of scribe 614 who is also present in the SW Area, in the Archives Complex and in Court 47 (Court 47). It is therefore a real chain reaction that inevitably leads to the date of the Pylos Linear B archives being traced back to a time on the border between the end of Late Helladic IIIA2 and the beginning of Late Helladic IIIB1 (Godart 2021, 91).

Conclusions

In conclusion, I would like to recall:

1) that Popham (1991) demonstrates that there was a re-occupation of the site of Pylos after the fall of the Mycenaean palace because vases from the Late Helladic IIIC or the very beginning of the Bronze Age were erroneously attributed to the final catastrophe which, in reality as the re-examination of ceramics suggests, dates back to the end of Late Helladic IIIA2 or the very beginning of LH IIIB1 (1300 BC);[8]

2) that the absence of ceramic forms typical of the Late Helladic IIIB2 at Pylos and the isolation of the site from the Argolis culture of the second half of the 13th century (between 1250 and 1200) suggest, as Thomas (2004) points out, that the palace was destroyed well before the Late Helladic IIIB2;

3) that the very high date associated with all the seals and seal impressions discovered in the destruction layer of Pylos does not fit well with a dating of the latter to Late Helladic IIIB2/IIIC (1200–1180 BC);

4) that the close similarities between the decoration of the throne rooms of Knossos and Pylos make difficult a hiatus of nearly two centuries between the destruction of these two sets;

5) that the ivories of Pylos, as demonstrated by Poursat (1977), are linked to a period much earlier than the Late Helladic IIIB1/IIIC and present elements, such as meanders and foliate bands, that recall the peculiarities detected in the frescoes: it is therefore tempting to associate the date of the frescoes and therefore the fall of the palace with the date of the ivories presenting this type of decoration;

6) that contacts were close between Crete and Pylos to the point that Cretan craftsmen were present in the palace of Nestor;

7) that the absence of amphorae with stirrup jars bearing inscriptions in Linear B at Pylos, while all the other continental palaces have restored in abundance this kind of objects dating from the end of LH IIIB1, suggests that the palace was destroyed before 1250 BC and that the contacts that Pylos had with Crete mainly concerned Knossos and;

8) that the belonging to the hand of the same scribe of tablets **La 994**, a document dated Late Helladic IIIA, and tablets **La 632** and **635**, as well as the connection between fragment **Xn 1449** dated Late Helladic IIIA and tablet **Vn 1339** attributed to Late Helladic IIIB2/IIIC, suggest that the fall of the palace of Nestor occurred at the extreme end of the 14th century or, better, at the very beginning of the 13th century BC.[9]

The end of Pylos would therefore have preceded that of the other palaces. It is likely that, alarmed by the catastrophe that had struck one of the centres of Mycenaean power, the authorities of Mycenae, Tiryns, Thebes and other continental palatial centres decided to build or strengthen the defence systems that protected their residences. It would therefore be from this time (extreme end of the 14th–beginning of the 13th century) that the first upheavals would have manifested themselves that led to the disappearance of the Mycenaean palatial civilisation.

I am pleased to insist on an essential point that has been largely ignored so far: the vases that are described in the series **Ta** of Pylos are exactly the vases with representations of animals on handles similar to those unearthed by Yannis Tzedakis in his excavations of Western Crete, in particular the double vase with a hare's head and papyri decoration of the Late Minoan IIIA2 (Godart and Tzedakis 1992, pl. LII). This allows us, in my opinion, to specify even more the date at which it is appropriate to locate the **Ta** series and therefore the archives and the fall of Pylos: Late Helladic IIIA2/IIIB1 at the very beginning, in other words the period between 1320 and 1300 BC (Godart and Tzedakis 1992, 22).

This period of the end of the 14th and the very beginning of the 13th centuries BC pre-dates the time (Late Minoan IIIB1=1250 BC) which sees the centre of Κυδωνία export its stirrups jars bearing inscriptions in Linear B towards the great palatial centres of the continent. This is the reason why there is no trace of these inscribed stirrup jars in Pylos.

The 'city' of Armenoi and Pylos

A tripod vase from a tomb of Volimidia, decorated with three animal heads (two deer, one bull) provides a good illustration of what scribe 602 describes in **Ta 641.1**. This vase is dated by Koehl (2006, nos 44–61) to Late Helladic IIIB1. Vases with animal protomes are rare but there is a series in the workshop of Kydonia dating from the Late Minoan IIIA2 (1370–1300 BC), which corresponds and corroborates the mention of the 'Cretan manufacture' indicated in the tablet. In addition, a vase with a handle, similar to the one discovered in the chamber tomb at Kalami but in this case certainly made in Armenoi, was unearthed in the Necropolis (R.M. 6600, see Postscript below).

This undoubtedly implies that there were relations between Pylos, Kydonia and, most important, with the 'city' of Armenoi at the end of the 14th century BC Indeed all these vases seem to pre-date 1250, or even 1300. If, as Poursat still writes to me: 'scribe 602, author of the **Ta** series, recorded material from his time, the latter could in no way be later than HR IIIB1' (Poursat, pers. comm. 20 June 2020), then I add that the comparisons between the descriptions of scribe 602 in **Ta**, the vases with animal's heads from the workshop of Kydonia, the vase of Armenoi and the vase from Volimidia argue favour of a dating located at the hinge between the end of the 14th and the beginning of the 13th centuries for the archives of Pylos (between the end of Late Helladic IIIA2 and the beginning of Late Helladic IIIB1). This conclusion is consistent with

that of Popham (1991) who, based on the re-examination of Pylos ceramics, suggests that the final catastrophe that swept away the palace of Knossos dates back to the period between the end of Late Helladic IIIA2 and the beginning of Late Helladic IIIB1, roughly equivalent to (straddles) Late Minoan IIIA2 and Late Minoan IIIB.

Postscript: the 'city' of Armenoi, *da-*22-to*

Yannis Tzedakis and Holley Martlew

Tomb 198

Tomb 198, that held the remains of two sisters or a mother and daughter (Chapters 4 and 5), contained R.M. 6600, a miniature double vase with an animal protome (see Fig. 6.19), made in an Armenoi workshop and dated Late Minoan IIIA:2. This find could be interpreted as additional evidence of a relationship and trade, between the 'city' of Armenoi with Kydonia, and with Pylos, and Mycenaean cities on the Mainland (see discussions above and below).

Tomb 159, the royal tomb

R.M. 9422, a human/male figurine (bust) who wears a full hat reminiscent of the white crown of Upper Egypt, as found in Tomb 159 (see Fig. 9.33). Because of a tear along the base the bust does not appear to be free standing. What is it? The figurine was re-examined by Tzedakis in July 2023. He reported it does not belong to a double vase, that it is indeed the bust of a man and it was definitely the production of a local Armenoi workshop. As it was found in the royal tomb, he suggests that 'With great imagination', it could have been the upper part of a sceptre … *wa-na-ka*, representative of the leader.

The clay bust is distinctive. Its eyes are wide and outlined; its eyes and nose are applied separately. With its stylistic similarity to the double vase with hare protome (Fig. 10.9) that Tzedakis found in the chamber tomb at Kalami (situated on the main road at the foot of Aptera, about 12 km from Rethymnon), it is surely by the same hand.

Figure 10.9. Hare protome on a double vase from a chamber tomb at Kalami.

With regard to the **Ta** series which contains descriptions of vases with animal heads that are said to come from the workshop of Kydonia, including the tripod vase from the chamber tomb at Volimidia, does this mean that vases with animal protomes could have been produced at more than one site (fashion), but would a potter with a definitive style (R.M. 9422 and the double vase with hare protome from Kalami) have carried out assignments in two workshops, Armenoi and Kydonia? This seems doubtful, unless the potter responsible was itinerant. That possibility raises a host of other questions. This is a complicated issue which is now the subject of further research. All vases with animal protomes, including the similarly dated **Ta** series (said) to come from the workshop of Kydonia and the tripod vase from Volimidia will need to be examined. As far as is presently known R.M. 9422 is the only known ceramic protome of a human being.

Whatever is revealed in future research however, it will almost certainly underpin evidence of a relationship and trade between the 'city' of Armenoi and Pylos, and with other Mycenaean cities on the Mainland. Most important of all, whatever is revealed concerning 1) the Late Minoan III double vases with animal heads (that were traded to the Mainland) and 2) the potter who developed his own style of representing the faces of animals on the handle(s) of double vases and on the face of a man, a 'ruler', on a diminutive ceramic bust, will serve to strengthen Godart's interpretation that the 'city' of Armenoi was **da-*22-to.**

Notes

1   If the palace of Pylos was never protected by a perimeter wall, the latter has no cyclopean character of the defense systems possessed by Mycenae, Tiryns, Thebes, Gla, Midea, etc (Zangger *et al.* 1997, 606–13).

2   Late Minoan IIIA2 is the time of major changes in the architecture of the houses discovered on the mound of Kastelli at Kydonia. See Godart and Tzedakis (1992, pl. LII) for the ritual hare-headed vase and papyri decoration of that time.

3   *ka-si-ko-no* in the series **Ra** of Knossos which accounts for the recording of swords is put on the same footing as *pi-ri-je-te*, an agent name in –τηρ; it is possible that both terms refer to craftsmen involved in sword-making; see Aura Jorro, 1985, *S.V.*

4   The presence of iconographic themes recurring in frescoes from different palatial sites suggested imagining the existence in the Aegean of the 2nd millennium, of itinerant painters who went from one palace to another and even, if we take into consideration the frescoes of Avaris which are undoubtedly the work of Minoan painters working in Egypt, from one country to another, to practice their art. This helps to explain the extraordinary existing similarity between the depictions of wild boar hunting in the frescoes of Tiryns and Orchomenos, and processions of Mycenae, Pylos, Tiryns and Thebes.

5   Popham (1991, 322) clearly sets the question: 'If we consider that the vessels of the Late IIIC period cannot belong to the destruction layer of the palace, then what is the answer to the question I asked myself: when was the palace destroyed? I would say: to a very ancient phase of the Late Helladic IIIB. This would explain the presence of the few decorated bowls and the conical kylix. This would also explain the strongly marked character "Helladic Late IIIA" of much of the ceramics, decorated or not. Among the vases cited by Blegen (p. 421) there is a stirrup jar imported from Crete that could perfectly belong to a considerably ancient period of the sequence in question'.

6   With regard to **Xn 1449**, the fragment was connected by Jose L. Melena to the tablet **Vn 1339** from Room 99 of the Palace (Godart and Sacconi 2020, 281).

7   Poursat (2014, 150) notes 'that the function of the South-West Building, very poorly preserved, remains undetermined; It does not seem to have been a first palace but rather a residential building contemporary with the palace'.

8   It is also the opinion of de Lis (2016, 533) who stresses: 'the high number of vessels that are stylistically earlier than the commonly accepted destruction date of the site, i.e., around 1200 B.C.'.

9   There is no doubt that The Mycenaean Palace from Aghios Vasileios which, like Pylos, is not defended by walls such as those of Mycenae, Tiryns, Midea or Thebes, looks very much like the palace of Nestor. While waiting for more precise information on the excavation of this great Mycenaean centre of Laconia, we must be satisfied with the preliminary remarks that the excavators have given us. According to the latter, the western 'Stoa' associated with the tablets unearthed on the site, would date from the Late Helladic IIIB1: 'So far, the West Stoa is undoubtedly the most important ... Its upper Storey contained several pithoi and a Linear B archive, the first ever found in Laconia. The Stoa was destroyed by an immense fire during the Late Helladic IIIB1 period, while the whole site was abandoned in early Late Helladic IIIC and it was reinhabited almost two thousand years later during the Byzantine times' (Karadimas *et al.* 2019). The date of the fall of Aghios Vasileios (Late Helladic IIIB1) would therefore be practically contemporary with the fall of Nestor's palace as I envisage it here (at the hinge between the end of Late Helladic IIIA2 and the beginning of LH IIIB1, around 1300 BC). It is exactly this dating proposed by E. Kardamaki (2017, 75): 'The following study aims to establish the sequence of pottery phases of the site from the early Mycenaean period up to the conflagration that destroyed the palace at the end of the 14th or at the beginning of the 13th century BC.' It was therefore in the south of the Peloponnese, in Messinia and in Laconia that the first turbulences would have manifested themselves, heralding the disasters which, towards the end of the 13th century, swept away the Mycenaean palatial civilisation.

Bibliography

Aura Jorro, F. (1985) *Diccionario Micénico* I. Madrid, CSIC.

Aura Jorro, F. (1993) *Diccionario Micénico* II. Madrid, CSIC.

Bittel, K. (1976) *Les Hittites*, Paris, Gallimard.

Blegen, C.W. and Rawson, M. (1966) *The Palace of Nestor at Pylos in Western Messenia* I. Princeton NJ, Princeton University Press.

Blegen, C.W. and Rawson, M. (1967) *A Guide to the Palace of Nestor*. Cincinnati OH, University of Cincinnati.

Blegen, C.W., Rawson, M, Taylour, L.W. and Donovan, W.P. (1973) *The Palace of Nestor at Pylos in Western Messenia* III. Princeton NJ, Princeton University Press.

Chantraine, P. (1968) *Dictionnaire étymologique de la langue grecque* I. Paris, Klincksieck.

Del Freo, M. (2016) I find-spot e la cronologia dei documenti in lineare B. In M. Del Freo and M. Perna (eds), *Manuale di epigrafia micenea. Introduzione allo studio dei testi in lineare B* I, 185–197. Padova, libreriauniversitaria.it edizioni.

Godart, L. (2020) *Da Minosse a Omero*. Torino, Einaudi.

Godart, L. (2021) *Les scribes de Pylos*. Rome-Pisa, Biblioteca di Pasiphae 13.

Godart, L. and Sacconi, A. (2017) *Supplemento al corpus delle iscrizioni vascolari in lineare B*. Rome. Biblioteca di Pasiphae 11.

Godart, L. and Sacconi, A. (2019) *Les archives du roi Nestor. Corpus des inscriptions en linéaire B de Pylos* I. Rome-Pisa, Pasiphae 13.

Godart, L. and Sacconi, A. (2020) *Les archives du roi Nestor. Corpus des inscriptions en linéaire B de Pylos* II. Paris, Pasiphae 14.

Godart, L. and Tzedakis, Y. (1992) *Témoignages archéologiques et épigraphiques en Crète occidentale du Néolithique au Mionoen Récent IIIB*. Rome, Incunabula Graeca 93.

Hiller, S. (1966) Knossos and Pylos. A case of special relationship? *Cretan Studies* 5, 73–83.

Karadimas. N., Vasilogamvrou, A. and Kardamaki, E. (2019) Preliminary remarks on the stratigraphy of the West Stoa from the new Mycenaean palace at Ayios Vasileios, Laconia. Unpublished paper presented to the Middle Helladic and Late Helladic Laconia. Competing principalities? conference, Athens, 12–13 April 2019

Kardamaki, E. (2017) The Late helladic IIB to IIIA2 Pottery Sequence from the Mycenaean Palace at Ayios Vasileios, Laconia *Archaeologia Austriaca* 101, 73–142.

Koehl, R.B. (2006) *Aegean Bronze Age Rhyta*, Philadelphia PA, INSTAP Academic Press.

Lis, B. (2016) A foreign potter in the Pylian Kingdom? A reanalysis of the ceramic assemblage of Room 60 in the Palace of Nestor at Pylos. *Hesperia* 85, 491–536.

Melena, J.L. (2000–2001) 24 Joins and quasi-joins of fragments in the Linear B tablets from Pylos. *Minos* 35–6, 357–69.

Palaima, T.G. (1988) *The Scribes of Pylos*. Rome, Incunabula Graeca 87.

Popham, M.R. (1991) Pylos: reflections on the date of its destruction and on its Iron Age reoccupation. Oxford Journal of Archaeology 10, 315–24.

Poursat, J.-C. (1977) *Ivoires mycéniens. Essai sur la formation d'un art mycénien*, Paris, Bibliothèque des Écoles françaises d'Athènes et de Rome 230.

Poursat, J.C. (2014) *L'art égéen 2. Mycènes et le monde mycénien*. Paris, Picard.

Shank, E. (2007) Throne Room Griffins from Pylos and Knossos. In P.P. Betancourt, M.C. Nelson and H. Williams (eds), *Krinoi kai limenes: studies in honor of Joseph and Maria Shaw*, 159–65. Philadelphia PA, INSTAP Academic Press.

Shelmerdine, C.W. (1998) Where do we go from here? And how can the Linear B tablets help us get there? In R. Laffineur and W.-D. Niemeier (eds), *The Aegean and the Orient in the second millennium: Proceedings of the 50th Anniversary Symposium, Cincinnati, 18–20 April 1997*, 291–9. Leuven, Aegaeum 18.

Skelton, C. (2009) Re-Examining the Pylos Megaron Tablets. *Kadmos* 48, 107–23.

Skelton, C. (2011) A Look at early Mycenaean textile Administration in the Pylos Megaron Tablets. *Kadmos* 50, 101–21.

Thomas, P.M. (2004) Some observations on the 'Zygouries' Kylix and Late Helladic IIIB chronology. *Hesperia Supplements* 33, 207–24.

Ventris, M. and Chadwick, J. (1973) *Documents in Mycenaean Greek* (2nd edn). Cambridge, Cambridge University Press.

Zangger, E., Timpson, M.E., Yazvenko, S.B., Kuhnke, F. and Knauss, J. (1997) The Pylos regional Archaeological Project, Part II, Landscape evolution and site preservation. *Hesperia* 66(4), 549–641.

Index